The Pig Iron Aristocracy

The Triumph of American Protectionism

Quentin R. Skrabec, Jr., Ph.D.

HERITAGE BOOKS
2008

HERITAGE BOOKS
AN IMPRINT OF HERITAGE BOOKS, INC.

Books, CDs, and more—Worldwide

For our listing of thousands of titles see our website
at
www.HeritageBooks.com

Published 2008 by
HERITAGE BOOKS, INC.
Publishing Division
100 Railroad Ave. #104
Westminster, Maryland 21157

Copyright © 2008 Quentin R. Skrabec, Jr., Ph.D.

Other books by the author:
The Boys of Braddock: Andrew Carnegie and the Men Who Changed Industrial History

All rights reserved. No part of this book may be reproduced or transmitted in any form or by any means, electronic or mechanical, including photocopying, recording or by any information storage and retrieval system without written permission from the author, except for the inclusion of brief quotations in a review.

International Standard Book Numbers
Paperbound: 978-0-7884-4515-6
Clothbound: 978-0-7884-7462-0

*To Our Lady of Damascus and the Iron Saints:
Dunstan, Eloi, Clement, and Leonard*

CONTENTS

Introduction vii
Chapter One: Gift from the Stars 1
Chapter Two: American Beginnings 9
Chapter Three: Genesis 19
Chapter Four: Early Beginnings..35
Chapter Five: Political Rise...55
Chapter Six: Rise of the Pig Iron Politicians......................73
Chapter Seven: The "Iron Whigs" 87
Chapter Eight: King Coal..103
Chapter Nine: Wheeling..113
Chapter Ten: Blast Furnaces 119
Chapter Eleven: War Demand...131
Chapter Twelve: Battle of the Furnaces 141
Chapter Thirteen: The Peak and Decline 149
Chapter Fourteen: Legacy 159
Bibliography..161
Index..165

Introduction

The story of pig iron is a personal one for me as a metallurgist, former steel mill manager, Pittsburgher, and an American. But from childhood, I was bombarded with the name of steel baron Andrew Carnegie. I slowly came to understand there were ironmasters long before Carnegie. Even in Pittsburgh there were the forgotten "old ones." But it was during a more recent trip to Niles, Ohio, while I was working on a biography of William McKinley, that I came to know the now forgotten "Pig Iron Aristocracy." It was at this Niles memorial to President William McKinley that I found my Pittsburgh roots. It was there in these overlooked bronze busts that I found many Pittsburghers as well as Ohioans that composed what the politicos of the time called the "Pig Iron Aristocracy." This started me on a search for America's early masters of iron who had built the foundation for men like Carnegie. This book is a result of research into my earlier biographies on George Westinghouse, William McKinley, H. J. Heinz, Michael Owens, and a group of Carnegie executives known as the "Boys of Braddock." The Pig Iron Aristocrats were romantic, colorful and bigger than life.

The Iron Age is synonymous with the Industrial Revolution in most countries. Until the information revolution, iron was a symbol of personal and national power. The knowledge of ironworking was held secret as a means to control that power. For centuries, the Philistines had prevented the Israelites from gaining that knowledge. Similarly, England tried the same thing with its American colonies. Without the ability to process iron, it would have been impossible to gain economic and political freedom. The Scots-Irish, in the hills of

Pennsylvania, Maryland, and Virginia, were the ones who opened the furnaces and forges of freedom. With iron, the colonies could stand up to the British, but even as the colonies gained their freedom, the British moved to economic warfare by limiting American iron production. A century later in Europe, Communists would come to the same revelation: control of the iron industry meant control over the people. For almost two thousand years, those who controlled iron production controlled the money, jobs, the economy, the military and political power.

The Pig Iron Aristocracy was a politically and industry-bonded group of American capitalists from the 1700s to 1900. Pig iron preceded steel (a product of pig iron) by over two hundred years as a manufactured material. The greatest accomplishments of the Pig Iron Aristocracy were the political rise of Henry Clay, the "American System," the formation of the Whig Party, the election of Abraham Lincoln, the formation of the Republican Party, the rise of American industrial power, the creation of American capitalism, the 1890 Tariff, and the election of William McKinley. The Pig Iron Aristocracy's golden years were from the 1850s to 1896, their peak being the McKinley Tariff of 1890. The Pig Iron Aristocrats had the largest political influence in the country, but they worked quietly behind the scenes. Political brokers such as Mark Hanna and Joseph Butler Jr. functioned as distributors of the vast wealth of these Pig Iron Aristocrats. In return, they were given representation in Congress and the Oval Office. They had many great senators in their camp such as Matthew Quay, James Ross, and Henry Baldwin, as well as house representatives such as William McKinley, James Garfield, and "Pig Iron" Kelly. They helped deliver the electoral advantage to Presidents such as Lincoln, Hayes, Garfield, Harrison, McKinley, and Taft. Their decline started with their greatest accomplishment: the election of President William McKinley.

The rise of the steel trust and powerful bankers such as J. P. Morgan would end the economic success of the old Pig Iron Aristocrats. With the passing of the Pig Iron Aristocracy's political might, the political power moved from the factories to banking centers and large corporate boards. Their legacy would be

America's industrial might and American cities, such as Pittsburgh, Youngstown, Canton, Niles, Warren, Wheeling, Cleveland, Akron, and Chicago. The decline of the Pig Iron Aristocrats began with the development of the integrated steel mill at Braddock, Pennsylvania, in 1880. With the integrated blast furnace, pig iron was no longer needed as an immediate raw material for steel. The end came with the formation of United States Steel and other steel corporations in the early 1900s. Only a few of the old aristocrats made the transition to steel and its large integrated corporations. As they had taken the torch from the iron "plantation" owners of colonial times, the Pig Iron Aristocrats passed it on to the steel barons.

The reader may well ask, What difference does the passing of power from wealthy pig iron manufacturers to corporate boards make? They found a way to blend parts of free trade and scientific protectionism into a form of capitalism known as the "American System." They resurrected the beliefs of Alexander Hamilton and Henry Clay to take America to industrial leadership of the world. The Pig Iron Aristocracy represented an alliance between labor and capital that promoted a national growth that went far beyond the obligations of a corporate board to the stockholders and bankers. The Pig Iron Aristocrats were capitalists and greedy, but they were nationalists. Their factories were in the major cities of America, in which they lived. Pig iron production was regional but had a national focus from the start. The Scots-Irish pig iron furnaces were in the hills of Pennsylvania, Maryland, Virginia, West Virginia, and Ohio prior to the Revolutionary War. Even in the era of the first blast furnaces, pig iron was a national enterprise. Although they would export, they were happy to control and dominate the American market for pig iron. The manufacture of pig iron formed the core of America's industrial centers. When the controls of the American iron and steel industry passed to New York capitalists and bankers, the focus turned to international money making. These new capitalists no longer honored the pledge of investing profit into American plant infrastructure, and politicians like William McKinley could no longer trust them to invest the rewards of protectionism. The Pig Iron Aristocrats were American aristocrats.

They lived in the communities where their furnaces were, and their mansions were discolored like the houses of the workers. Their investment capital came from local banks and built local furnaces and mills. Their capitalist greed was limited by their passionate love of this country. Pittsburgh was once the capital of the Pig Iron Aristocracy, but their final strongholds were Youngstown and Niles, Ohio.

CHAPTER ONE

Iron: Gift from the Stars

Iron has been the foundation of national wealth, not from its intrinsic value, but in its manufacture. It is in the manufacture of iron that cities rise and nations form. For millennia, iron has been the measure of civilization. The metallurgical knowledge needed to release it from its ore has been cherished and often kept secret.

The source of all the iron on earth and in the universe is the nuclear furnace of supernovae. Iron's tightly bound nucleus is the break-even point of fusion and fission within stars, and thus it is the end-point of nuclear creation. Iron is the heaviest element that can be produced in a star's nuclear fusion furnace. Iron is the end product of the star's nuclear reactions and heralds the death of the star in supernovae. The fiery explosions of supernovae inject iron into the universe. Even man's first experience with iron came from supernovae-propelled meteorites. These meteorites, rich in iron and iron family elements, came to earth ready for manufacture. The oldest iron artifacts are Egyptian from around 3500 B.C. Examples of meteoritic iron implements were found in the form of knives and axes in King Tut's tomb.

Iron has a much more important role than its use as a strategic metal. It is fundamental to our very existence. Iron is essential to almost all life on this planet. Its best known role is to carry oxygen from the lungs to organs such as the heart and brain, and on the return trip it returns the waste gas, carbon dioxide, to be exhaled.

Seventeenth-century physician, Vincenzo Menghini, discovered this through experiments on dogs. Early in the eighteenth century, physicians linked iron, or the lack of it, to anemia and fatigue. Eighteenth-century pharmacists dispensed sugar-coated iron pills to help heal and improve general health. Many also carried a tonic of iron and wine. Generally, most civilized countries supply a diet of 20 milligrams, which is more than the 7-11 milligrams recommended. Many countries like the United States require iron additions in baked goods. Still, it is estimated that over 500 million people suffer from anemia due to low iron.

Prior to the discovery of its medical role, iron played a mystical role in many civilizations. The Egyptians and Greeks considered iron a gift from the gods. Iron also represented the power of the gods such as Mars, Vulcan, and Thor. Medieval farmers, however, took the power of iron to new heights. Iron was believed to protect one against demons and witches. The iron horseshoe, in particular, was displayed to protect houses. Blacksmiths formed special guilds around the mystical powers of iron. Blacksmith practices even had spiritual components. A blacksmith would hit the anvil every fourth time to keep the devil in chains. The blacksmiths were one of the earliest trades to have patron saints. St. Clement (the fourth pope) was the general patron, but farriers (horse-shoers) selected St. Dunstan. Smaller guilds selected their own patrons, such as St. Eloi and St. Quentin. Many shrines of the saints developed a ritual of iron offerings, such as St. Leonard in Germany. On St. Leonard's Day, believers inundate the shrine with horseshoes and iron chains. Ultimately, Leonard became the patron saint of prisoners because of the iron chains found all over his shrine. These mystical roots formed the heritage of Germany's Iron Cross award.

Iron is the most abundant element of the earth, yet it is much more difficult to extract than copper. For centuries, space remained man's only source. The earliest recorded history shows the importance and precious nature of this meteoritic iron. Homer's classic, the *Iliad*, written around 800 B.C. notes that one of the prizes for winning an athletic contest was a large lump of iron. The Old Testament is full of references to iron. In Genesis, Tubalcain is mentioned as being a blacksmith. Tubalcain, born a seventh generation from Adam, was described as "an instructor of every artificer in brass and iron." The Book of Jeremiah has numerous references to the processing and manufacture of iron, suggesting Jeremiah was a metallurgist. Supernovae explosions were the source of all this early iron. It represented an unusual metal in an early world of copper and bronze. Iron was harder, giving a basic military advantage to its possessor. Its inherent hardness allowed it to hold an edge, so vital to victory in hand-to-hand combat. The early civilizations of Iraq evolved into world powers based on their knowledge of iron. Iron remains the winning edge in close battle. It is a lesson we have again found as we retool vehicles with armor in today's Iraq.

Being in the "iron club," like today's "nuclear club," represented membership in the ancient superpowers. This exclusive club included the Hittites, Assyrians, Philistines, Chaldeans, and Babylonians. Hittites are generally given credit historically for discovering the how to free iron form its ore, although recent findings suggest an African origin. The Hittites built a pre-Christian empire based on their knowledge and use of iron. We have seen that the Philistines ruled the Fertile Crescent with their iron advantage. The Greeks had ruled centuries before, using bronze technology, which they passed on to the Romans. Their empire, however, was dependent on developing their iron technology, since iron swords are considerably stronger than bronze. It would be like a mechanized army going against light infantry. Even to this day, the knowledge and understanding of iron metallurgy is a measure of a nation's military might.

Iron ore is the result of millions of years of stellar iron oxidation. A complex metallurgical reaction is required to remove iron from its ore. Carbon has a great affinity to oxygen and is used to take (reduce) oxygen from iron oxide ore. There are a number of ores (iron oxides) of commercial value. Magnetite is a dark heavy oxide of seventy-two percent iron, making it magnetic. Hematite represents the majority of the ore available. It takes its name from the Greek "haima," meaning blood, because of its reddish brown color, although it can have a black appearance. Hematite is seventy percent iron and is the basis of the Great Lakes ore deposits. Local deposits of iron ore are common throughout the world, but they are usually a low-grade type of bog or kidney ore.

Iron, as noted, cannot be readily taken from its ore without some knowledge of metallurgy, and the origin of that knowledge remains a mystery to this day. That knowledge was held in secrecy by smiths and even governments. In the later centuries before Christ, the Philistines and Israelites jointly shared the Jordan Valley. It was illegal for Israelites to work with iron or even use iron implements. It was also illegal for Philistines to discuss ironmaking with the Hebrews. The mystic nature of iron and the secrecy surrounding it probably is related to the prophet Jeremiah's fascination with iron and metallurgy.

The reduction of iron ore into iron requires the use of carbon, which can be supplied from wood, charcoal, coal, or coke. The use of carbon to reduce or take oxygen out of iron oxide ore requires two other things to drive the chemical reaction: heat and an oxide flux. These things complicated the discovery of ironmaking and allowed the Bronze Age to blossom. The earliest combination of these chemical factors seems to have occurred in Africa maybe as early as 2000 B.C. The procedure consisted of a male ritual of drinking and ironmaking. The drinking ritual seems to remain part of the process to this very day. Iron ore was gathered and put in a large dug pit. Some limestone and wood were added. The males lighted the fire and would spend days and nights blowing air into the pit through long blowpipes to raise the temperature. After days of hard work and drinking, the pit was uncovered to find a mass of high carbon iron at the bottom of the pit. This lump of high carbon iron would require more "heating and beating" to make a usable piece of steel for knives. The lump today would be called pig iron or cast iron because it contained up to 3% carbon. Heating and beating would remove carbon, so the smith could turn it into flexible steel, which has a carbon level well under 1%. Continued heating, hammering, and beating could produce a very low carbon product (under 0.02% carbon) known as wrought iron. Actually the pasty mass of the African furnace had lots of impurities besides excess carbon because the air blowing produced a low heat compared to later improvements such as bellows.

Iron was a king maker in ancient times. The Hittites had an iron monopoly until about 1000 B.C. The Hittites had discovered the mixing of charcoal, ore, and limestone with heat resulted in reducing iron from its oxide. The Hittites also tapped the power of bellows. A wood fire can reach 1650 degrees F; with blown air you can reach 2300 degrees F, which is needed for iron smelting. The Hittites were the source of the Celtic iron-making technology that spread through Europe. The Japanese had their own pig iron reduction method known as Tatara. This utilized the bowl-like basin of the Africans and Celts, but they added a four-foot stack to increase the power of the air blast. This Japanese method took furnace temperatures to new highs and produced an impurity-free iron. Actually they produced a high carbon steel called Wako, which was used for their famous Samurai swords.

Iron was made in very limited amounts until the seventeenth century, when high-stack charcoal furnaces started to be built. These stacks ranged from ten feet to fifty feet high. They were top-fed with alternating layers of charcoal, iron ore, and limestone. These charcoal furnaces were common in the United States from 1700 to the end of the Civil War. Old stacks can still be found in Western Pennsylvania, the Hocking Hills of Ohio, Western Maryland, and Northern Michigan. Some were even pressed back into service during World Wars I and II. Amazingly, some charcoal furnaces are still used today in the wood rich areas of South America and China. Today huge blast furnaces of coke, iron ore, and limestone are used to continuously produce pig iron.

Initially nations and kings controlled the manufacture and processing secrets of pig iron production. With the advent of the charcoal furnace and the spreading of iron metallurgy, iron production had a democratizing influence. Ironmaking passed into a craft in the seventeenth century. Carleton Coon, in *The Story of Man* described the democratization: "Craftsmen multiply until they cease to be tame purveyors to kings, and work primarily for the people.

Their standards of excellence and their price ceilings are not set by royal overseers, but by their own guild chiefs whom they themselves have elected. Thus during the Iron Age did a middle class grow big enough to produce its own institutions." Of course, the rise of the blast furnace would create a new type of capitalistic "royalty" known as the Pig Iron Aristocracy. These men would forge metallurgy and politics. The versatility of pig iron would change the nature of technology and science for centuries.

Pig iron or furnace cast iron is iron with about three percent carbon in it. As a liquid it flows readily and can be cast into shapes. As the iron-carbon product, cast or pig iron is a hard, strong, and wear-resistant product, but an extremely brittle product as well. The term pig iron refers to the casting of long blocks of cast iron, which looked like suckling pigs. The term pig iron came about from the casting system of charcoal furnaces in the 1700s. Cast iron has some unique properties as an engineering product as well. Benjamin Franklin invested in an iron furnace to produce stoves. Cast iron absorbs heat and distributes it evenly, making it a favorite material for stoves and frying pans. There was a large colonial market for cast iron kettles for the production of salt. It was also the preferred cannon metal by the 1800s, replacing brass. Cast iron stoves and products were often cast directly at the charcoal or blast furnace, but a furnace cast was limited to a small amount every twelve hours or so. Pigs would be stored cold and re-melted in bigger air furnaces to produce larger or great amounts of castings. Thus pigs became a standard product on the market for foundries and rolling operations. Pig iron was also the base material for the production of wrought iron and steel.

Wrought iron and steel made up another growing market for pig iron in the 1800s. Pig iron could be worked into wrought iron or steel by heating and beating. Wrought iron and steel offered a soft, malleable product that could be used for axes, hammers, knives, gun barrels, bayonets, sheet metal, horseshoes, tinplate, nails, and rails. Wrought iron was the stock working material for blacksmith shops. To make steel, a blacksmith would re-introduce carbon into the wrought iron to improve strength and durability. Early Celts learned and advanced the science and art of blacksmithing. By re-heating and hammering the high carbon of the pig or furnace iron is removed. High removal of carbon results in soft and usable wrought iron. A more controlled lower removal rate would result in the amazing combination of strength and ductility known as steel. By controlling the precise level of carbon, a smith could control the final properties of the steel.

Steel may seem like a strange alloy of iron, since it actually has less carbon in it then cast iron, but more than wrought iron. Plain carbon steel is in fact, closer to elemental iron than cast iron. The removal or control of carbon results in the stronger and tougher material we know as steel. Steel, like iron, changed the political maps because of its superior strength and workability. The processing of high quality steel, like its parent material, remained tightly held secrets in places like Damascus, Toledo, and Sheffield. Early in history, steel had been the product of a smith beating hot "iron" on an anvil to remove carbon. The heat and beat method was great for farriers, blacksmiths, and sword makers, but could not support the needs of the Industrial Revolution. Pittsburgh metallurgist William Kelly invented the process, Henry Bessemer got credit for it, and Andrew Carnegie commercialized it. The Age of Steel represents the end of the Industrial Revolution Pig iron as an individual and separate product out-produced steel tonnage until the late 1890s. Tools came into a new era with the dawn of steel. Alloy steel gave us the automobile.

Early blacksmiths and sword makers experimented with carbon levels and properties via the anvil. They learned that higher carbon levels gave a higher hardness. The also discovered that hammering burned carbon out. To add carbon, they put the steel into a pile of hot charcoal, thus they learned to add and take away carbon. Sword makers even learned to fold and layer steel by using the anvil and hammer. Even today, good steel knives that are etched show this layering. This etched layering is the distinctive mark of Damascus steel. Sword makers also learned to impart a low carbon core to the sword for flexibility, while layering a high carbon surface for hardness and holding an edge. Japanese sword makers took this to an art form. Sword makers also learned to "quench" steel by religious ceremonies that drove the hot sword into an animal, using its body fluids to quench the steel to hardness.

    Sheffield, Toledo, and Damascus had been the ancient steel centers, but the rise of the Industrial Revolution gave rise to new cities of steel, like Essen, Pittsburgh, and Youngstown. While Toledo, Sheffield, and Damascus depended on the production of charcoal iron and then the "heating and beating" to produce steel, these new centers made steel in a very direct means. The eighteenth- and early nineteenth-century process consisted a high stack furnace using charcoal, limestone, and air bellows to produce high carbon pig iron. The pig iron was then taken to a Bessemer converter where air was blown through the molten pig iron, burning out the carbon. The length of the air blow controlled the amount of carbon in the steel and thus the properties desired for its end use. A result of serendipity was that Bessemer steel was high in sulfides, which gave it good machining properties.

CHAPTER TWO

American Beginnings

When iron implements were needed for farming, fishing, and shipbuilding, Puritan entrepreneurs jumped in. With high-quality bog iron ore nearby, the Puritans tried their hand at iron manufacture. John Winthrop Jr., son of the governor, went to London to raise capital to build an ironworks. Winthrop had been trained in metallurgy and chemistry at Trinity College, and is truly America's first industrialist. Winthrop got some capital but opened the investment up for colonists as well. A number of Puritan ministers invested. By 1645, two ironworks were in operation around the Boston area. Saugus was America's first large industrial complex, having an iron furnace, rolling mill, and slitting mill, but it struggled for capital throughout its existence. The Pilgrims also learned the technology and started operations by the late 1600s. The New England example spread to other colonies until the British tried to restrict colonial iron implement production with the Iron Prohibition Act of 1750. The idea was to force the colonists to ship pig iron to England to be converted into product and shipped back as taxed product.

The pig iron industry was one of America's first stimuli for industrialists to come to America. By the late 1600s, England had lumbered out the hardwood that fueled their iron furnaces. What hardwood that was left in the British Isles was reserved for strategic shipbuilding. Britain became dependent on Swedish pig iron to make stoves, wagon tires, farming implements, and firearms. The

Swedish iron, because of its low impurity levels, even today is considered the best in the world. As a superintendent of an electric furnace for tool steel production, I was purchasing Swedish iron in the 1980s! The problem in the late seventeenth century was the poor political relationship between Britain and Sweden. In the early 1700s, British agents were sent to America to set up "iron plantations." One of these agents was James Farmer, who in 1717 set up an iron furnace on Principio Creek, Maryland. Farmer was part of a British-backed firm. He had water power, hardwood, and iron ore, but like most of these early plantations, he lacked labor.

Plantation managers tried using Native American laborers but they would run away, and black slaves were considered too expensive. The plan was to use British indentured workmen and convicts who had ironmaking experience. Commonly these indentured workmen came for five years and were cheaper than paying fifty pounds for a black slave. Some of England's best managers and furnace men were sent to run the operations. In 1717, strained relations with Sweden caused a boom in American iron production. Colonial governments further encouraged iron production. These encouragements included land and other special exemptions. For example, all taxable inhabitants in Maryland were required to labor on highway building, but iron plantation workers were exempt. In 1724, James Farmer and his Principio Furnace would make the first pig iron shipment to England.

Virginia Governor Alexander Spotswood brought seventy Palatinate Germans to Virginia to start an ironworks at Spotsylvania and Germania. Spotswood added a hundred slaves to cut wood for blast furnace fuel. This group of expert German ironworkers would be important in the early development of iron plantations. It was an expensive operation that failed because of poor ore and high overhead costs. The Germania operation mined ore and cast it into pigs at the headwaters of the Rappahannock River, then shipped the pigs fifteen miles to be remelted and cast into products such as firebacks and andirons. Spotswood sent agents into the Blue Mountains of western Pennsylvania to set up more furnaces. He did map what would become the Arundel iron ore formation, that runs through Baltimore to Frederick, and then on to the Potomac River. Within years, furnaces spotted the Arundel trace through Maryland and Virginia. Many of the Germans found work in other American furnaces. Spotswood also granted Captain Augustine Washington (George's father) an ironmaking operation on the Potomac River. One problem with both Virginia and Maryland iron ore was its high level of sulfur, which made pigs hard to process into iron products. Washington became an owner of Principio Company. In 1767 the furnaces of the company were producing 17,000 tons of pig. Some 3,000 tons were shipped to England, but it was known to be of poor quality.

Besides Virginia, the iron industry was flourishing in New Jersey and Pennsylvania. One of the first great pig iron masters was Peter Grubbs, who opened Cornwall Furnace in eastern Pennsylvania's Lebanon County in 1734. In the 1740s, his furnace was producing thirty tons of pig iron a week. Grubbs expanded to hammer and forge operation early on. The Grubbs family maintained the operation for decades, supplying the manufacturing "bloom" town of Lancaster, Pennsylvania. The Cornwall Furnace closed in 1883 after 141 years of continuous operation. Other furnaces such as Elizabeth Furnace arose in the 1750s to further supply the needs of Lancaster. One of those furnace men, George Ross, was a signer of the Declaration of Independence. Furnaces also opened in the 1750s in New York and New England to supply the growing metal industry in the Connecticut Valley.

Even with limited efforts to produce pig iron, the Connecticut colonists developed a secondary metals industry that was unequaled in the other colonies. Connecticut developed eight rolling mills (two more than the total of all the other colonies), which manufactured nails, hoops for barrels, and sheet for tinplate. The demand for iron bar created a search for bog iron, which was found in 1734 in Salisbury Township near the New York border. Initially, these industrious colonists used crude Catalan furnaces and small forges to make iron. By 1740, charcoal furnaces were being built; some of these became "iron plantations," making an array of products. Connecticut ironsmiths advanced the technique of steel making for small tools, knives, and bayonets. The area became known as the "Arsenal of the Revolution." In 1762, Ethan Allen, future Revolutionary War hero, built his Lakeville Furnace. He added some of the large forges in America to make cannons and anchors. Without the Salisbury iron works, it is doubtful that the colonists could have armed themselves properly against the British. In the in 1880s, the heritage of the iron works of the Connecticut Valley made the valley part of the Pig Iron Aristocracy politically.

Predominantly agricultural, colonial America supplied most of the world's raw cotton and tobacco, but it also supplied one-seventh of the world's iron. The northern states excelled in the production of iron because the South lacked good iron ore deposits. Over two hundred furnaces operated in Pennsylvania alone during the 1700s. This was known as the era of iron plantations, which tried to mimic their Southern tobacco cousins. As noted, the Scots-Irish pioneered American iron production with cash coming from English partners. Three presidents, George Washington, Abraham Lincoln, and William McKinley can trace their roots to northern colonial ironworking families. The northern iron production tried early on to use the plantation system that had been used in Northern Ireland. Scots-Irish freedom loving and Quaker moral values prevented slavery from taking root. The heart of the early iron production was in the Schuylkill, Susquehanna, Juniata, and Monongahela River valleys. Iron production required water power, iron ore, and large amounts of hardwood forest. The stimulus for the growth of these iron plantations was the need for iron in England in the later half of the seventeenth century and the early part of the eighteenth century. The British, however, both needed and feared the rise of American iron production. Ideally, the British wanted to ship raw pig iron to England duty-free, then process it into product and ship it back to the colonies. The relationship reflected the heavy British investment in colonial iron works, as well as a strategic need. England's major source of iron for its military needs was Sweden, which allied itself with France during the 1730s. The Americans more and more started to produce products for their own market, resulting in decades of economic controls imposed by the British. Scots-Irish owners, in particular, wanted freedom from British oversight, which had been the problem for them in Northern Ireland.

The early American furnaces were known as charcoal furnaces, since charcoal was the fuel. Charcoal was produced from the incomplete burning of hardwood. Workers known as "colliers" managed the process. The word, "collier," came from the word "coal," meaning the wood was turned to "coal." A large furnace required as many as twelve colliers and a master collier. A collier piled the burning wood and covered it, so it would convert to charcoal through incomplete combustion. It took ten days for a smoldering pile to convert wood to charcoal, and required constant tending by colliers. The collier was considered a type of craftsman, and early furnaces had an apprentice system.

A typical charcoal furnace was thirty feet high and made of limestone blocks. A charcoal furnace required huge amounts of wood to operate. Larger eighteenth-century furnaces could consume 840 bushels or twenty-two cords of wood in a twenty-four hour period. That translated into almost an acre of wood! A rough estimate for a charcoal furnace was a half-acre of wood per ton of iron produced. The average furnace turned out twenty to twenty-five tons a week. It's no wonder that by the start of the eighteenth century, Britain's hardwood had been lumbered out by iron production. Britain had to pass laws to restrict the use of oak, in particular, because it was a strategic material for the shipbuilding industry. Hickory was considered the best for charcoal, but oak, chestnut, and walnut were commonly used. America's huge hardwood forests kept furnaces on wood into the mid-1800s, while England was forced to switch to coal a hundred years prior. It is interesting to know that even today Brazil still uses charcoal, and its use is creating a lumbering out of wood and rain forests there. Wood gave America a significant cost advantage over British iron. On the iron plantation, colliers were wage earners, although indentured whites and freed Negroes often did woodcutting, and in some cases Negro slaves. Slavery was legal in Pennsylvania until 1780, but was limited in most cases to iron plantations because of the Quakers' anti-slavery beliefs. Free workers earned a wage based on productivity, that is, on how many cords of wood cut. Again this productivity wage system was uniquely American, although the British had incorporated it into some factories. Where slaves were used on an iron plantation, the "task" system was mostly used. Slaves performed the necessary furnace tasks, and then were free to farm and hunt for their family. Slaves had their own family cabin, rifles, and a plot of land to garden. The task system had major economic advantages over the slave system of Southern plantations. Iron plantations in Maryland and Alabama used the task system until the Civil War. The plantation master got the work free and did not have to carry the burden of food and board. Slave productivity was significantly higher as well with this injection of capitalism into the slavery system. Still, it was morally wrong and opposed by the Pennsylvania Quakers.

The furnace crew was made up of a manager, two founders, two keepers, two guttermen, and three to four laborers or "fillers." Wages were paid on productivity based on tons produced. Furnace workers in Pennsylvania were a mix of Scots-Irish, German, Welsh, and English. Germans and English were commonly brought in by the Scots-Irish owners because of their technical expertise. The furnace ran twenty-four hours a day for about nine months, at which time the furnace was rebuilt or lined for a new "blast." The operation required twelve-hour shifts, which would become an industrial standard in America. Physically, the furnace was usually built into a hill, so the furnace could be charged easily with wagonloads and wheelbarrows of iron ore and charcoal. They were built on streams or "runs," so water power could drive the bellows for an air blast. The ore was cast into raw "pigs" for further processing, but kettles, pots, fire plates, and stoves were often cast directly. These cast products required craftsmen skilled at molding (founders), in which the Germans excelled. Germans were actively recruited during the eighteenth century. Experienced miners were needed also, and the Welsh and Cornish were highly recruited. Miners were also paid on tons dug, instead of by the hour.

Workers and their families lived on the plantation. The close community of Germans and Anglo-Americans forced the Germans to adopt English as their language, and also led to some intermarrying. Each family was given a plot of land to work, but women and children helped in many other aspects of plantation manufacturing. Plantations often grew cash crops, such as corn, rye, and hemp. From these crops, they raised hogs and sheep, distilled whiskey, and made rope. Plantations often had gristmills and smoke houses. Distilling of whiskey almost went hand–in-hand with ironmaking. Drinking was a constant problem for these iron plantations. The ironmaster maintained a "company store" where goods could be purchased cheaply by the workers. More importantly, the iron furnace was just part of a large industrial complex that included rolling and slitting mills, forges, and blacksmith shops. Forges were needed to break down the cast iron into a softer iron product. That "bloom" could then be rolled and slit in nails. The bloomer, forge, and rolling mill jobs required less skilled workers, and employed more Negroes, both slave and free. The apprentice system did allow even slaves to work their way up the organization. Labor shortages plagued the industry throughout the colonial period.

The iron plantations' greatest growth was from 1700 to 1776. The great iron plantations of eastern Pennsylvania were part of the migration of ironmasters from New England, New York, and New Jersey. During that period, twenty-two iron furnaces, forty-five forges, four bloomeries, six steel operations, three slitting mills, two plate mills, and one wire mill were built in Pennsylvania. Most of these iron plantations were in eastern Pennsylvania, and a few families, such as the Colemans, Birds, Putts, and Grubbs controlled the pig iron industry. The ironmasters of the plantations were the most colorful of the lot and lived like royalty. In the post-Revolutionary War years, the pig iron industry moved west, and so did the new iron families such as the Measons and Shoenbergers. These owners lived in huge plantation houses similar to those of the Southern planters. These furnaces started out casting implements and stoves, but moved to an immediate pig iron product that could be rolled or forged to products. The iron plantations evolved into more integrated factories. They developed a romantic tradition in the American culture of the northern states. Pig iron was a strategic war material, and therefore had great economic value. As in past history, these ironmasters controlled the economy, trade infrastructure, and political systems. Government became dependent on these iron works as a source of defense and power. Even ironworkers were the royalty of the working class, demanding top wages and social status. They were often exempt from military service.

The Iron Prohibition Act of 1750 played a major role in the start of the American Revolution. England would allow pig iron production, but it had to be shipped to England for processing. Foundries, rolling mills, wrought iron manufacture, nail making, and slitting mills were all prohibited. Finished iron products were then taxed. The ban united the iron plantation owners and manufacturers as well. Owners looked to central Pennsylvania and other backwoods counties to avoid the taxes on iron. Often Scots-Irish iron pioneers started small furnaces in the hills of central Pennsylvania, Maryland, Virginia, and Kentucky to produce wrought iron. Existing mills were excluded from the ban but taxes and tariffs were used to restrict their growth. It is no wonder that many patriots, such as George Washington, Nathanael Greene, Paul Revere, Charles Carroll, Arthur St. Clair, Ethan Allen, George Ross, and James Irwin descended from iron plantation families.

The first notice of iron ore in western Pennsylvania came in 1780, based on a survey of the Monongahela River. In the 1790s, richer ore deposits were found in western Pennsylvania, which brought in ironmasters from Maryland, Virginia, and eastern Pennsylvania. Besides ore, western Pennsylvania offered virgin hardwood for fuel. The first of these western Pennsylvania furnaces was Alliance Furnace at Jacob's Creek, Fayette County, Pennsylvania. The furnace was the investment of the merchant firm of William Turnbull and Peter Marmie of Philadelphia. Scots-Irish Turnbull supplied the capital and Frenchman Peter Marmie the technology. This firm of Turnbull and Marmie was part of the international Scots-Irish trading network that would be the seed of Pittsburgh and Youngstown's pig iron industries. Many of the first Pig Iron Aristocrats were connected to the firm of Turnbull and Marmie.

Many of these first American industrialists failed because of labor shortages, but they built a future for America. There were six signers of the Declaration of Independence, several Revolutionary War generals, and many colonels. At least seven became judges and one was a Supreme Court judge selected by Washington. The ironmasters were from the middle class, much like their Southern counterparts, and some had risen from furnace workers. They were a mix of English, Welsh, German, Scots-Irish, and Scottish. The families actually celebrated intermarriage with their "iron marriages." The ironmasters lived in mansion houses on 8,000 to 10,000 acre plantations. Eastern Pennsylvania mansion houses were older and more equivalent to Southern plantations. Western Pennsylvania had much smaller (usually Scots-Irish) iron furnace operations. Ironmasters sometimes had slave servants. During the colonial and post-Revolutionary War period, most ironmasters struggled due to lack of labor and capital. Still, the shortage of capital allowed for workers to invest their savings into the operation, something uncommon in Europe. Furthermore, as workers gained knowledge and money, they often moved further west to open their own iron furnace operations, or progressed to a master craftsman with new opportunities. The peak years would be 1800 to 1845, when iron plantations gave way to iron mills. What is often overlooked is that America's iron self-sufficiency allowed it to wage and win a war with Britain. Connecticut ironworks alone produced 2,000 light cannons. Cannonballs were made throughout Pennsylvania and Virginia. The Pennsylvanian rolling mills supplied colonial gunsmiths with wrought iron bar for barrels.

CHAPTER THREE

Genesis of an Industrial Culture

The Pig Iron Aristocrats included immigrants with German, British, and Scots-Irish roots, but the Scots-Irish dominated the industry. They forged a blend of religion and capitalism that made them the most productive of the colonial ironmakers. The Scots-Irish heavily settled the iron hills of Pennsylvania, Maryland, Ohio, and Virginia. They combined whiskey and ironmaking with fur trading. They had ties to a vast trading network that could export pig iron with shipments of tobacco and furs. Just as important was their hatred of the British, which resulting in defying iron production bans. The Scots-Irish need to be understood, to understand the Pig Iron Aristocracy.

The term Scots-Irish consisted of a loose amalgamation of the Ulster Scots-Irish, lowland Scots, Presbyterians, and a mix of Protestant Scots and Irish in America.[1] Some early Irish Catholics even took to calling themselves Scots-Irish to avoid discrimination. The Scots-Irish dominated the colonial frontier and the formation of the "American System" in all aspects. Fifteen American presidents have claimed Scots-Irish ancestry, including three of pure Ulster Presbyterian lineage: Andrew Jackson, James Buchanan, and Chester Arthur. Philosophers and leaders included Alexander Hamilton, Patrick Henry, James Madison, John Calhoun, Daniel Boone, Sam Houston, and twenty-one signers of the Declaration of

---

[1] The term Scots-Irish is the most inclusive, although the Scots-Irish are specifically the Ulster Presbyterian Scots.

Independence. The Scots-Irish excelled in all phases of trade and American industry, claiming the likes of Thomas Mellon, Andrew Carnegie, and Henry Ford. They started and dominated industries such as pig iron, steel, whiskey, and wool. Especially in business, these Scots-Irish had an informal clan of Scottish reciprocity going back to frontier days. The Scots-Irish dominated the American trade distribution of furs, tobacco, cotton, and pig iron. Few realize that Glasgow, not London, controlled colonial trade. The Scots-Irish, more than any other immigrant group, are responsible for the rise of the Pig Iron Aristocracy and industrial America.

The roots of the American Scots-Irish run deep in the history of Europe. They learned trading and plantation management the hard way. The Protestant Reformation not only split Europe, but also divided many countries such as Ireland and Scotland. The struggle emerged on old historical lines, such as the Gaelic highlanders, Episcopal highlanders, and the Calvinist lowlanders. The Calvinist Church, known as the Kirk, evolved into the political and religious foundation of nationalist Scotland (and the American Presbyterian Church). England tried to break both the Scottish Kirk and the Irish Catholic Church, which would create a new "nation" in Ireland of Ulster Protestant Scots. Scots were encouraged with land to move to Ulster, Ireland, to create plantations and suppress the Irish Catholics. The Ulster Scots had hard times in Scotland, and they wanted more opportunity.

The Ulster Scots not only wanted religious freedom but economic freedom. In many ways their move to Ireland was the first step to America. Scotland's lowlanders had become world traders in the 1600s, controlling the New World tobacco market and the American fur trade. This trading network and mercantile system would give birth to a type of economic democracy. Scottish lowlander Adam Smith would herald that new approach in his book, *The Wealth of Nations*. Along with these economic ideas, the Scottish enlightenment would develop democratic concepts such as the "pursuit of happiness" that would be incorporated in the writings of Thomas Paine, Thomas Jefferson, and James Madison. More than any other nation, Scotland integrated the ideas of economic freedom, democracy, and world trade. And these Scots-Irish would bring these ideas to America.

Ulster in Northern Ireland had traditionally been its own community, geographically separated by mountains. The Gaelic Irish of the area had close ties with the Scottish Gaelic highlanders (only twenty miles of ocean separated the two), while the mountains often made them more distant from the southern Irish. The mountainous environs of Northern Ireland and Scotland are similar to that of the Appalachian Mountains of Pennsylvania, Maryland, Kentucky, and Tennessee, in which the Scots-Irish would settle. The British conquest of Ireland in the 1500s had meant stiff resistance in Northern Ireland around Ulster. The Gaelic Irish had continued an insurgency for decades until a union of Scotland and England subdued them. These Gaelic Catholics were stripped of their land and rights.

England wanted to "colonize" Ireland and eliminate the Catholic faith. England opened the counties around Ulster to Protestant Scots, barring Catholic ownership, and thus restricting most of the Catholic Scot highlanders from emigrating. At the same time the union of Scotland and England had brought the ideas of the English church into the Scots Calvinist Kirk, which was the heart of the lowlanders. These Scottish Calvinists started to flood into Ulster between 1608 and 1618, to gain freedom of religion. The English saw the Scots Kirk as the lesser of two evils. The English tolerated it as a means to move out the Irish Catholics. Protestant lowlanders of Scotland started to make the move. Estimates suggest that as many as forty thousand Scots came to Ulster between 1608 and 1689 as the religious struggles between England and the Scottish Kirk continued. Ulster became a Scottish fortress and the protectorate of the Kirk. These Ulster Scots hated the English as much as the Irish Catholics. The Ulster Scots were never prosperous, but they did make a living off the land, wove linen, developed blacksmithing skills, learned how to use coal as fuel, and learned iron smelting. The Scots-Irish became renowned for their mastery of the Celtic art of blacksmithing, which originally brought many to the American colonies. Just as important, they learned world trading.

Pennsylvania had originally been founded by William Penn's English Quakers, but was soon joined by the Germans searching for more religious tolerance. In the 1710s, Penn sent agents to recruit the Scots-Irish tenant farmers of Northern Ireland. Penn's agent offered them cheap land, and religious and political freedom in return for settling the expanding frontier. These Ulster Scots were ready for a new opportunity to advance economically, and an opportunity to get away from the interference of the English church. The Scots saw advantages in going to Pennsylvania over the formally sanctioned Southern colonies, which had a more tropical farming environment. The Scottish and Scots-Irish immigrant stream split in the 1750s, with some moving out of central Pennsylvania to the Carolinas and another branch going west. All headed for mountainous areas, settling in the "hollows," or along "runs." Men like John Fraser established frontier trading and blacksmithing posts. George Washington visited Fraser's post on the Monongahela in 1757 when he first explored the "Ohio Country." In fact, the Scots-Irish pushed the frontier every year, running from government and taxes. They left a few members at their frontier villages, which were often backfilled with German settlers.

The Scots-Irish pushed the frontier, led by traders such as John Campbell and George Croghan. By the 1750s, Croghan had formed an extensive and lucrative trading network that stretched into the Great Lakes. The British invested and supported this Scots-Irish network because it competed directly with the French in the New World. The great trade warehouses of Glasgow, Scotland, were major distributors of furs to Europe. This trade network would bring the founder of the Pig Iron Aristocracy, James O'Hara, to Fort Pitt.

The Scots-Irish developed frontier industries around their trade network such as ironmaking and whiskey production. They first developed these industries in the Cumberland Valley of Maryland, then moved to western Pennsylvania and in the early 1800s to Ohio.

They built large charcoal iron furnaces to manufacture a variety of frontier products. These smaller Scots-Irish iron furnaces would compete with the larger iron plantations of central and eastern Pennsylvania. As the Scots-Irish moved down the Monongahela River Valley towards Ohio, they established America's whiskey industry. Wagon roads were lined with Scots-Irish blacksmiths to repair wagons and shoe horses. These early rest stops also produced and sold rye whiskey. This "Monongahela Rye" won international acclaim as it was shipped to Europe via the Scots-Irish trade network. It was the Scots-Irish whose village became today's Pittsburgh, and even today the linguistic accent of Pittsburghers is the mark of the Scots-Irish.

During one of the most trying periods of the Revolutionary War, George Washington noted: "Should it come to the worst, I will fall back into the mountain region of Pennsylvania and make my stand among the Scots-Irish there." Washington had worked with the Scots-Irish of the Ohio Company and knew well their toughness and perseverance. In those same trying years, General Washington depended on the patriotic army chaplain (a Scots-Irishman from western Pennsylvania), Henry Brackenridge, to hold the army together. A volunteer group of 130 Scots-Irish was the first to come to the relief of the Continental Army at Boston in 1775. Perhaps more importantly, it was the charcoal iron furnaces of the Scots-Irish that supplied cannonball to the Continental Army. The Scots-Irish loved freedom and independence more than their other American counterparts, which arose from their natural and long hatred of British control and taxes.

In the 1790s, the new American nation experienced the backbone of the Scots-Irish in their resistance to a federal tax on whiskey. By the 1770s, the Scots-Irish of western Pennsylvania had prospered, owning farms and dominating the little village of Pittsburgh. As one floated down the Monongahela River in 1790, the smoke of stills filled the surrounding hills (an estimated twelve thousand stills!). Whiskey production was an old Scottish art that now flourished in America. Monongahela Rye had gained a reputation as on a par with Scotch whiskey. As you floated to the confluence of the Monongahela and Allegheny, you could often run into flotilla of keelboats taking Monongahela Rye to the port of New Orleans. From there Monongahela Rye was shipped to the East coast and into the European trading network via Scotland.

The animal life of the Ohio Country or "Allegheny Plateau" reflected the rich diversity of the vegetation. Large herds of wood buffalo, elk, and deer roamed the area in the 1600s. Wolves, panthers, wildcats, and bears were a common sight as well. Streams (called "runs" by the early Scot-Irish) and the stream valleys (known as "hollows" by the early Scot-Irish) were home to beavers, muskrat, foxes, and many species of fish. Now extinct, passenger pigeons darkened the skies for hours at time. Turkeys roamed throughout the area. John Audubon painted his famous rendition of the passenger pigeon in this primal countryside, and the area remained a bastion for hunters of this species into the 1870s. The now extinct Carolina Parrots also were abundant. Poisonous copperheads and rattlesnakes were numerous, but the early Scot-Irish seemed to fear the boundless mosquitoes more. By far the species of mammal that most left its mark on western Pennsylvania was the Scots-Irish of Ulster.

The Scots-Irish were best represented by the immigrants of Presbyterian Ulster Scots, who a hundred years earlier had migrated to Ulster, Ireland, the west coast of Scotland being about thirty miles away. Western Pennsylvania today remains the best linguistic legacy of the Scots-Irish with strong accent of the population and the unique vocabulary, such as hollows, burghs, and runs. The accent of western Pennsylvania combines the burr of the Scots with the brogue of the Irish and adds the gutturals of Germany. Many words and expressions remain in the local language, such as "yinz" or "younz" for you or you people. Other words include "redd up" for clean up and "slippy" for slippery and "still" for steel. Street names of the area still include the Scottish term "diamond" for town squares. Pittsburgh's center today has a Diamond Street.

The dominance of the Monongahela Valley by the Scots-Irish had been a slow process with deep roots in Scottish and English history. The term Scots-Irish is a mix of Scots-Irish, Scots, and later combinations of Scots and Irish in the colonies. Scots-Irish is uniquely American with no such term existing in Ireland or Scotland. In fact, prior to 1830, the term "Irish" generally meant Scots-Irish. Surveys of western Pennsylvania suggest a very complex mix indeed, and very reflective of America's melting pot. There remains much debate as to the makeup of Scots-Irish. On the frontier, the Scots-Irish of Ulster readily mixed with German, English, Scots, and Irish, usually pulling them into the Presbyterian Church. While at times considered uncivilized by urban dwellers, they were the "upper class" of the frontier. Many nationalities seemed willing to assimilate into a Scots-Irish family.

By 1730, the Presbyterian Scots-Irish dominated the trade and agriculture of western Pennsylvania. By the Revolution, the Scots-Irish dominated the politics and were the earliest recruits to the Continental Army. Driven by the deep belief in Calvinistic destiny, the Scots-Irish were the nation builders. They had an inherent hatred of the British and taxation going back to their days in Ulster. They represented the extremes of beliefs. Drinking was considered a human right, while any work on Sunday was a major sin. On the Ohio frontier they often represent the industrial elite as while as the hillbilly lower class. The new American government soon learned their hatred for taxes in the 1790 Whiskey Rebellion, which was the first challenge to the federal government. The Scots-Irish were fiercely independent, which naturally drew them to the wild frontier areas. While many left for Kentucky and Tennessee from the Monongahela Valley, many stayed to build banking and manufacturing empires. The values of the Presbyterian Ulster Irish would constrain Pittsburgh business into the 1870s, not allowing Sunday work.

Both Philadelphia and Norfolk took on large populations of Ulster Scots and Scots, but most of them pushed inland to the west and south. Scots immigration was a chain process. The Ulster Scots led the way, often followed by Scottish highlanders and lowlanders, who developed settlements and the Presbyterian Church as the Ulster Scots moved out. Many Ulster Scots had ties going back centuries to the lowland Scots, and the Scots-Irish quickly aligned with the lowlanders in Scotland for an international trade network in their American settlements. The majority of the Ulstermen headed to central Pennsylvania, then on to western Pennsylvania. The Ulster Scots led the Scottish spear of immigration inland, but found what they were searching for in the hills of western Pennsylvania, West Virginia, Eastern Ohio, Maryland, and Kentucky. The Ulstermen were poor farmers who preferred fur trade and whiskey production. By the 1750s, the Scots-Irish Ulstermen dominated the valleys of the Monongahela, Allegheny, and Ohio rivers commercially. Even the loyal British colonists of Virginia depended on the Scots-Irish to hold back the French, who were migrating from the Great Lakes region into the Ohio and Monongahela valleys.

The Scots-Irish formed cooperative communities with the frontier Indians. This is somewhat surprising considering some of bloody battles that arose over the years. The Indians and Scots-Irish seemed to have a natural affinity. They were economically united in trade. Both were restless subsistence farmers who preferred to be traders. They seemed to agree with the communal view of land cherished by the Indians. They both had a love of whiskey. They were free spirits with a dislike for formal government and, both liked the formation of clan-type communities versus urban centers. They were both fighters and warriors by nature. The Scots-Irish demonstrated a similar symbiotic relationship with German settlers.

By 1750, the Indians, Scots-Irish, and lowland Scots formed a powerful fur trade empire. In 1752, over fifty percent of Pennsylvania exports to London were furs.[2] Even in the agricultural states such as South Carolina, fur comprised ten to fifteen percent of the exports. The Scots controlled the fur trade and the supply-chain logistics to Europe. The Ulster Scots hunted and traded with the Indians for furs. The Scots Presbyterians reinforced the supply chain with a string of settlements across Pennsylvania that could move furs from the west. The furs were assembled and transported to the

---

[2] Jane Merritt, *At The Crossroads* (Chapel Hill: University of North Carolina Press, 2003).

seaports by trading networks of Scots-Irish such as George Croghan, who had wagon trains and riverboats throughout the colonies. Croghan utilized the many Scots settlements to get his wagon trains across the Appalachian Mountains. Croghan represented the major logistics chain from the fur country to American ports, with Scots also controlling the ports. Men like Croghan and Daniel Boone expanded into the ginseng trade through the fur trade network.

The Scots had the largest fleet of trading vessels in the 1700s. Early in the 1700s, they became the carriers of colonial tobacco, with 386 ships engaged in the trade. In fact, by 1745, the Scottish imports of tobacco from America exceeded that of London and all English ports. Scottish shipping dominated the ports of New York, Philadelphia, Alexandria, and Baltimore. The control of tobacco shipping was well over fifty percent, giving the Scots domination of the Virginia colony.

A great deal of the political power of the Scots-Irish (as the colonists began to call them) came as their role as a Rosetta Stone for Indian communications and trade. Pennsylvania employed them as Indian agents and the Virginians used them for trade. The most important of the Scots-Irish frontier politicians was George Croghan, who would be Pennsylvania's Indian agent for the decades preceding the American Revolution. Croghan became the "King of Traders," and his Scots-Irish trading network rivaled that of all of New France. By the late 1740s, his network not only included the Pennsylvania frontier but most of present-day Ohio with key trading posts on the Miami, Monongahela, Sandusky, Ohio, Walsh, and Cuyahoga Rivers. By early 1750, Croghan was pushing up against the French in Detroit and the northern Great Lakes. Some of these outposts were the farthest points of white colonization, deep in the wild and well beyond any forts or settlements.

Croghan played a balancing act throughout the 1750s in Monongahela country. While Croghan started out as a Pennsylvania agent, he was soon drawn to his Scottish brethren in Virginia and Governor Dinwiddie (who governed from 1751 to 1757). Dinwiddie was a Scotsman and Virginia had argued early on that the

Monongahela Valley and west was outside the Pennsylvania charter. Dinwiddie was from the Glasgow trading district that had supported the Scottish Act of Union in 1707. Prior to Dinwiddie's arrival, a group of Virginians including Lawrence Washington and Thomas Lee had formed the Ohio Company to settle western Pennsylvania and develop trade with the Indians. Upon his arrival, Dinwiddie became a major shareholder, seeing it as a means of participating in the lucrative Scottish fur trade. He already had financial interests in Glasgow banking, warehousing, and shipping. Croghan also saw the Ohio Company as a means as getting military support, as the aggressive French traders were moving into the Allegheny, Ohio and Monongahela valleys in the 1750s. Croghan would eventually become part of the Scots-Irish trading empire of John Campbell. Another Scots-Irishman, John Fraser, set up a blacksmith shop at the junction of Turtle Creek and the Monongahela River in 1749 near Braddock (at today's USS Edgar Thomson Works). A young area explorer, George Washington, visited Fraser in 1752.

The conflict with the French came to a head in 1752 as the French pushed south into the Allegheny River Valley, overrunning many of Croghan's trading posts as well as those of other Scots-Irish traders, such as John Fraser and John Findlay. Croghan lost posts and employees throughout the Allegheny and Ohio valleys as well as in Kentucky. John Fraser retreated to his home on the confluence of Turtle Creek and the Monongahela River while Croghan reinforced his headquarters on the lower Allegheny. By the summer of 1753, the Scots-Irish fur trade was shut down. This threat to the Scots-Irish trading empire would change the world. Virginia's Governor Dinwiddie immediately realized the threat to the Scottish and British fur trade by the French advance. He dispatched twenty-one-year-old George Washington with frontiersman Christopher Gist to take a letter of concern to the French in western Pennsylvania in late 1753. On November 23, Washington met with Scots-Irish traders at Fraser's Turtle Creek home and later with Croghan at Logstown. After delivering the letter, Washington returned to Williamsburg to inform Dinwiddie of the French defiance. Dinwiddie dispatched a regiment of forty-one

Virginians to build a fort at the confluence of the Allegheny and Monongahela where the Ohio River formed. The Virginians teamed up with local Scots-Irish to build the first encampment at the site of today's Pittsburgh, as George Croghan and John Fraser had suggested in the 1740s.

During the French and Indian Wars, England and the colonies would unite. Then England sent General Edward Braddock and two "Irish" regiments: Sir Peter Halket's 44th Foot and Thomas Dunbar's 48th Foot. These regiments had a mixed makeup, probably the majority being Scots highlanders and Scots-Irish. Officially, the 48th was listed as forty percent English, ten percent Scot, thirty-four percent Irish, and sixteen percent American.[3] With Roman Catholics being banned from the army, the listed Irish had to be Scots-Irish, Protestant Irish-English, and a considerable number of undeclared Roman Catholics. The poor conditions in the Scottish highlands in the 1750s made it a recruiting gold mine for the stretched British Empire. Once in America, many of these would desert to blend in with the Scots-Irish population. Washington's Virginians and the Carolina Americans would join Braddock's 44th and 48th. It should be noted that these units were considered the British's poorest as opposed to the famous Highlander regiments such as the "Black Watch."

The road that Braddock built followed a wide trail developed for decades by Scots-Irish immigrants through the Cumberland Valley (originally an Indian path). On July 9, 1755, Braddock, with Halket's 44th, crossed the Monongahela at Fraser's trading post on Turtle Creek. The French and Indians ambushed the force, and the result was the bloodiest defeat of the British on American soil. The field that day would have more future generals than any other known fight, among whom were George Washington, Thomas Gage, Horiatio Gates, Daniel Morgan, Charles Lee, John St. Clair, Adam Stephen, James Craik, and James Burd. Other notables included Pontiac, the father of Chief Tecumseh, Daniel Boone,

---

[3] Stephen Brumwell, *Redcoats* (Cambridge: Cambridge University Press, 2002).

George Croghan, Christopher Gist, and William Shirley. Sir Peter Halket, who died that day, was a distant cousin of Scotsman Andrew Carnegie, who would build his first steel mill on this very site over one hundred years later. General Braddock also died in this defeat, and the town of Braddock honors him. The battle would change the nature of battles on American soil forever.

Braddock's defeat caused many of the frontiersmen abandon their settlements, while only the Scots-Irish remained at their reinforced trading posts. It changed British army recruiting as well. Lord Loudon was appointed commander of American forces after the death of Braddock. He listened to Washington and Croghan, and formed light ranger units of American frontiersmen. He rebuilt the British regiments with pure Scottish highlanders in place of lowlanders, poor Irish, and Englishmen. The Highlanders would lead England and General Wolfe to a major victory at Montreal. George Croghan and his company became the proprietors of the Indian and fur trade. With a shaky peace with the Indians and British security, yet a frontier lack of laws, the Scots-Irish started to settle in the area. At Fort Pitt they became tailors, merchants, and started light industries such as blacksmithing. Letters from soldiers and frontiersmen brought more Scottish immigrants to Fort Pitt and the Monongahela region. Its tradition of trade made Pittsburgh the world's largest inland port.

The backcountry farms were poor by European standards, but they produced crops such as flax, hemp, and rye. Backcountry industries were started by the Scots-Irish, which included weaving flax into linen and hemp into rope. Rye was used in place of Scotch barley to make whiskey, and Monongahela Rye was becoming world famous. The Ohio River allowed shipments of these products to New Orleans and then to the East Coast and Europe. Such trade led to the production of oak and hickory barrels as well as flatboat construction. By the 1770s, the Scots-Irish were permanent settlers, with some estimating that they were the second largest ethnic group after the English and ahead of the Germans. The term Scots-Irish now represented a broader population of Ulster Scots, Scottish,

Irish, and mixed blood. The Scots-Irish would make up more than one-third of the Revolutionary War soldiers.

The "Whiskey Rebellion" of the western Pennsylvania Scots-Irish in the 1790s would dig to the root of American domestic politics. The Monongahela Valley was filled with the smoke of whiskey stills in the 1790s. Rye whiskey was a mainstay of the area's Scots-Irish. As a British colony, the whiskey production had been controlled and taxed, but the remoteness of the Monongahela Valley made it almost impossible to collect the taxes. President Washington and Alexander Hamilton imposed an excise tax on whiskey in 1794. Scots-Irish General Neville was chosen as tax collector in western Pennsylvania, even through he had initially opposed the tax. The tax schedule varied but it was around six to ten cents a gallon (a gallon of rye whiskey sold for a dollar). The valley Scots-Irish mustered a militia and burnt the estate of General Neville. It was the first test of the federal government. A few days later, their leader, General James MacFarlane, was killed. For weeks Scots-Irish militia roamed the area. Hamilton persuaded Washington to send 13,000 troops to western Pennsylvania to put down the rebellion. Before the militia had reached Pittsburgh the uprising was defused as the Presbyterian Church preached enforcement of the law. Some Scots-Irish, however, moved into Kentucky and Tennessee to produce their whiskey. The new frontier was out of the reach of the federal tax collectors. These Pennsylvania emigrants would form the Kentucky and Tennessee bourbon and whiskey families of today. Another group of these industrious Scots-Irish would stay and move into pig iron production, which offered new opportunities. Some of the Scots-Irish plantations produced both whiskey and pig iron, and in the 1790s charcoal furnaces started to dot the Kentucky countryside.

The real heart of the Scots-Irish empire was the Presbyterian Church. The Presbyterian Scottish church evolved into a unique role, as did the Scottish order of Freemasons. Once the Ulster Scots cleared land and established a settlement, missionary pastors were sent from the Philadelphia Presbytery. The role of the pastor was both religious and social. After services, the Sabbath was for socializing. The Presbyterian Church commonly brought in Germans and some English. This social role continued in the twentieth century, making the church a favorite with capitalists and businessmen from many protestant backgrounds, such as Henry Clay Frick, Andrew Carnegie, George Westinghouse, and Thomas Mellon. Another role of the church was education. Presbyterian schools were the first in western Pennsylvania and supplied highly educated teachers to the frontier. The excellent educational opportunities and social events drew in Germans as well as Scots-Irish. The system would create many of the American leaders of the nineteenth century.

Ultimately, the Scots-Irish of the Ohio Frontier would create an American industrial empire led by men such as Andrew Mellon. This industrial valley would arm America in the War of 1812, the Civil War, World War One, and World War Two. Politically, the Monongahela Valley would change Jeffersonian agrarianism into Republican industrialism. The Republican Party would be founded there. The Ohio Frontier would become the bulwark of Ulster Presbyterianism. There is an old story that says John Knox prayed, "O! Lord give me Scotland," and God granted it, throwing in Pittsburgh for good measure. And the Scots-Irish made western Pennsylvania the heart of frontier higher education by establishing colleges such as the University of Pittsburgh, Washington and Jefferson College, Allegheny College, Geneva College, Grove City College, and Westminster College. The area brought forth musicians such as Stephen Foster and inventors such as Robert Fulton. Linguistically, the valley remains Scots-Irish even to this day, and culturally, the valley still prides itself on "worship on Sunday and whiskey on Monday." The valley has produced more steel than any other place on earth and probably has drunk more whiskey too. It is clearly the fatherland of the American Scots-Irish.

At the heart of the success of the Scots-Irish was a Calvinistic belief in self-destiny. These Scots-Irish settlers had worked hard and were farmers, plantation owners, manufacturers, brewers, weavers, and industrialists. The Presbyterian Church, which had forged the frontier identity of the Scots-Irish, evolved into a social and economic network for enterprising Scots-Irish. The Scots-Irish had formed the politics of America as well as the concept of pragmatic democracy. They had evolved from farmers to industrialists and international traders. Not surprisingly, they formed banks to support their enterprises, and this financial system in the early 1800s would be the basis for an industrial revolution in the late 1800s. The Scots-Irish bonded well with the German and English colonists to promote business and trade networks. Their iron furnaces of Pennsylvania and Ohio were the foundation of today's steel industry. In northeastern Ohio, the Scots-Irish mixed with Connecticut Yankees on what became Ohio's "Western Reserve." This mixture gave rise to America's pig iron industry. This "Pig-Iron Aristocracy" formed the Whig Party in these iron districts in the 1840s, and later helped give rise to Scots-Irish Republican President William McKinley.

CHAPTER FOUR

Early Beginnings

The rise of industry in the west starting in the 1790s fueled the engine of immigration. In the 1790s there were a number of rising western manufacturing centers such as Cincinnati in hog processing, St. Louis in fur trading, New Orleans in trade, Lexington in manufacturing, and Pittsburgh in manufacturing. Outside of New England and eastern Pennsylvania, a vast manufacturing area started to arise on the Allegheny plateau west of the mountains. It was referred to early on as the "Ohio Frontier" or "Ohio Country," but today it includes western Pennsylvania, all of western and southern Ohio, the panhandle of West Virginia, and parts of Maryland and Kentucky. The plateau was rich in wood, coal, iron ore, oil, limestone, and rye for whiskey, quality industrial clay, and water transportation networks. The area in the 1700s was roughly 40% Scots-Irish, 30% to 35% English, 10% Scots and Welsh, and 10% German. The Scots-Irish dominated the area, and tended to organize informal clans. The "capital" of the industrial plateau was Pittsburgh, and its main transportation routes were the Ohio River for the west, and the Cumberland Gap to the east.

This rich plateau's first mention was from George Washington's pre-Revolutionary War visits to the area. He was one of the earliest to note outcroppings of coal that could be used for heating. Many of the British generals of the Revolutionary War had planned to divide up this rich plateau after their victory. Many early Pittsburgh settlers had been Revolutionary War officers (and Scots-

Irish) including: General William Butler, General Presley Neville, General John Gibson, General John Wilkins, General Richard Butler, General Arthur St. Clair, Colonel John Irwin, Colonel James O'Hara, Colonel John Campbell, Colonel Ebenezer Denny, Colonel George Morgan, Colonel Isaac Meason, Colonel Thomas Cresap, Major Isaac Craig, Major Abraham Kirkpatrick, Major Stephan Bayard, and many others. These officers found land and trading connections including military supply contracts for Fort Pitt. The military supply connection continued into the 1830s, and clearly contributed to the area's industrial rise. Almost all of these founding citizens became part of the Pittsburgh's manufacturing elite known as the "Pig Iron Aristocracy." The Pig Iron Aristocracy was really part of the first industrial military complex of the nation.

The Scots-Irish had come to the area early on to avoid the taxes and regulations of government. By 1790, Pittsburgh was a major western shipbuilding area, a major user of heating coal, a cluster of forge shops and rolling mills to manufacture nails for the building boom, a fur tanning center, the whiskey production center of America, and a western trade center. No longer "hillbillies," The Scots-Irish moved into the city and became the founding fathers of Pittsburgh. In Pittsburgh, the Scots-Irish truly formed an industrial clan. This industrial clan allied with English-speaking Protestant Germans and English. The Pittsburgh clan of manufacturers had three membership "requirements"; they were members of Lodge 45 of Ancient York Masons, the U.S. Army, and the First Presbyterian Church. Lodge 45 actually used the Presbyterian Church as their meeting place. The Masonic Lodge was popular with the area's Scots-Irish and English, and probably was a key part of the business network. Lodge 45 offered a place for religious differences among Protestant groups to be put aside for business reasons. Another common organization among the Pig Iron Aristocracy was the U.S. Army. At least, initially the military bond was the strongest bond of all. Pittsburgh's manufacturing and trading for almost fifty years was centered on supplying the army.

The manufacturers of western Pennsylvania formed a new type of business network. Often through marriage they made ties across religious and ethnic lines. Partnerships across family businesses were also common. The pig iron manufacturers were also invested in glass, boatbuilding, brewing, and trade. These early Pig Iron Aristocrats tended to put business ties ahead of any of the "requirements." Later, Republican Party membership would also evolve as a requirement, almost all evolved from the Federalist and Whig Parties. The early founders of the First Presbyterian Church were Pittsburgh's key manufacturers, which included James O'Hara, Ebenezer Denny, John Wilkins, Isaac Craig, Hugh Henry Brackenridge, and Samuel Bayard, and again all of these were militia officers and most were Scots-Irish or related through marriage to Scots-Irish. One well researched study showed the founding families of Pittsburgh to be 68% Presbyterians.[4] Another common organization of these manufacturing barons was the Pittsburgh Library Company, a private library.

A 1790 survey of western Pennsylvania including the five western counties: Allegheny, Washington, Fayette, Westmoreland, and Bedford, supports the assumption of much intermixture of blood.[5] The survey of 12,955 families showed 37% English with the "Scots-Irish" making up 36%. Of the 36% Scots-Irish, there were 17% Scots, "Scots-Irish" were 7.5%, 2.7% were Ulster Irish, 4.7% were English-Irish, and 4.6% were southern (Catholic) Irish. The survey supports the blended view of the term "Scots-Irish." This early survey is surprising in that it refutes the common belief of Ulster Irish controlling the area, but they were on an equal par with the English based on numbers. The Scots-Irish, however, controlled the political and economic heart of the area. The reason for this perception was the dominance of the Presbyterian Church as a frontier church. The Presbyterian Church reflected the values and the hearts of the Ulster Irish. The Presbyterian Church bound frontier Protestants into a political force. The Presbyterian Church

---

[4] Joseph Rishel, *Founding Families of Pittsburgh* (Pittsburgh: University of Pittsburgh, 1990).
[5] Solon and Elizabeth Buck, *The Planting of Civilization in Western Pennsylvania* (Pittsburgh: University of Pittsburgh Press, 1967).

cannibalized other frontier Christians, particularly, the Protestant Irish, Scots, and English. Another usual part of the "Scots-Irish" legacy was that they were often found in Pennsylvania and Maryland alongside Protestant German settlers. The Germans not only assimilated with the Scots-Irish, but also claimed their identity to improve their social status.[6] Most frontier communities had Scots-Irish living side-by-side with Germans. Often as the Scots-Irish pushed west, the Germans built on Scots-Irish foundations. By the 1800s, the Presbyterian Church had taken in many Protestant Germans in western Pennsylvania as well.

Interestingly, even with the dominance of the Scots-Irish in "Ohio Country" there was no Presbyterian preacher in the area to after the Revolutionary War. One of the founders, John Wilkins reminisced about arriving in the area in 1783: "all sort of wickedness were carried to excess and there was no morality or regular order." Before the 1790s, most of the Scots-Irish lived in the hills, and were bound by no laws. The fort town of Pittsburgh was more a tavern stop than manufacturing center. The first of the elite Scots-Irish manufacturers might have been John Johnston, who was a clockmaker and possibly Pittsburgh's first manufacturer. His clock repair business on the frontier grew beyond his ability to get raw materials. He started a small wire mill and encouraged the building of the Anshultz charcoal iron furnace to supply his mill. The Johnston family would be considered one of Pittsburgh's founding families.

Perhaps the most important Scots-Irish pioneer on the Ohio frontier in the 1700s was John Campbell. Captain Campbell was a surveyor attached to Fort Pitt in the 1760s. Campbell is credited with laying out many of Pittsburgh's streets, such as Second Avenue, Market Street, Grant Street, and Water Streets. More importantly, Campbell worked to create a trading empire by bringing together a network of individual of Scots-Irish fur traders,

---

[6] Helen Vogt, *Westward of ye Laurall Hills* (Parsons: McClain Printing, 1976).

such as George Croghan, John Fraser, and Thomas Cresap. Campbell was part of the Ohio Company of Virginia. Campbell also had developed strong ties to the Penn family and Scots-Irish traders in Glasgow and Liverpool. In the late 1760s, Campbell teamed up with John Symonds to form the trading firm of Campbell and Symonds, which had a monopoly on English fur trading from Pittsburgh to Louisville. Campbell owned significant real estate in both Pittsburgh and Louisville. His firm traded not only furs, but also supplied merchandise to the early frontier. Campbell was connected to the then manufacturing center of Lancaster, Pennsylvania, allowing him to bring iron implements to Pittsburgh for sale. Lancaster preceded Pittsburgh, Cleveland, and Chicago as America's first manufacturing town.

Lancaster not only moved farm products to Philadelphia, but gathered furs and whiskey from the western frontier for shipments to Philadelphia through Campbell's firm. The German and Scots-Irish connection and network strengthened through the 1730s as wagons ventured west. Trade brought manufacture such as saddle making and gun making to Lancaster to supply the frontier. Lancaster as a growing manufactory started to attract not only German but also Swiss and Scots-Irish gunsmiths. In the 1730s, Swiss gunsmith Peter Leman started to develop the "Pennsylvania rifle" at his Lancaster shop. European smooth bore muskets were not suited for hunting in the American backwoods. Leman's new rifle had a longer barrel, improved sights, reduced bore, and better balance. It quickly became popular on the American frontier. By 1740, Lancaster was the capital of American gun manufacture and the home of our greatest gunsmiths. In 1752, General Braddock appointed a Lancaster Scots-Irish gunsmith as the master armorer to his army. In the 1770s, Lancaster was critical in supplying the colonial army with guns. Lancaster's population passed 10,000 with leather and clothing trades growing as well. In 1773 the city had fifteen master weavers, nine master stocking weavers, thirty shoemakers, ten tanners, seven saddlers, five skinners, and two boot makers.

German craftsmen at the western terminus of the road developed the "Conestoga Wagon" also called the prairie schooner to move freight. These giant freight wagons named after Conestoga Creek at Lancaster. These wagons were watertight and the wooden wheels were reinforced with iron belts. Teams of six to eight horses pulled these freight wagons. They moved in convoys of up to hundred wagons. By the 1740s, there were around ten thousand Conestoga wagons in use. The road branched into western Pennsylvania, Virginia and the Carolinas. Benjamin Franklin's mail system was extended through these wagon trains to Pittsburgh. In the 1750s, it was said that German Lancaster employed thirteen blacksmiths, five wheelwrights, twenty joiners, and seven turners in the production of these manufactured wagons.[7] The Germans bred "Conestoga" horses to further improve the overall transportation system. John Campbell was instrumental in tying this early manufacturing center Lancaster to Pittsburgh merchants. Goods could be moved to Pittsburgh in two weeks at about $120 per wagon load. By the 1760s, Pittsburgh started to manufacture much of its own goods, instead of importing them from Lancaster. In the 1780s, the firm of Campbell and Symonds would bring future capitalist James O'Hara to Pittsburgh. By the 1790s, Pittsburgh had replaced Lancaster as a manufacturing supply center for western expansion. The number of blacksmith shops in Pittsburgh increased as Scots-Irish smiths came in from Maryland and the Carolinas. These blacksmith shops were flourishing by western and military needs.

The roots of the Pig Iron Aristocracy go back to the original "clan" of Scots-Irish Pittsburgh manufacturers. The prominent founders were James O'Hara, Isaac Craig, and Joseph McClurg. These men started out as merchants, but invested their earnings into manufacturing. General Isaac Craig came to Pittsburgh during the Revolutionary War and later became commander of Fort Pitt. He stayed on in Pittsburgh as a real estate investor and merchant in the 1790s. Later he served under General James O'Hara as the army

---

[7] Carl Bridenbaugh, *The Colonial Craftsman* (New York: Dover Publications, 1990), 119.

quartermaster general. This partnership of O'Hara and Craig invested in shipbuilding and glassworks. Joseph McClurg started a foundry in 1804 to supply the military. The foundry was called Pittsburgh Foundry on Fifth and Smithfield Streets. This company would be known as Fort Pitt Foundry during the Civil War, and it was the predecessor of Mackintosh-Hemphill Company. The foundry produced cast iron cannon and howitzers for the army in the 1810s. In the War of 1812, it produced cannon and ball for both Commodore Perry on Lake Erie and General Andrew Jackson in New Orleans.

The rise of Pittsburgh's manufacturing was tied to the military connection of Scots-Irish Generals James O'Hara and Isaac Craig. It had helped supply Anthony Wayne in the 1790s. In the early 1800s, a supply arsenal was built by Colonel William Foster (the father of songwriter Stephen Foster). The arsenal created demand for pig iron to make cannon and cannonball. Foundries and blacksmith shops opened rapidly to support the military demand. Most of the pig iron was coming from the western Pennsylvania charcoal furnaces that were being opened to support the demand of the Pittsburgh foundries and blacksmith shops. Foundries re-melted pig iron using coal-fired air furnaces known as "cupolas." Pittsburgh had massive coal deposits across the Monongahela River on "Coal Hill," now known as Mount Washington. This coal seam, known as the "Pittsburgh Seam," also surfaced upriver at Braddock and Turtle Creek. The immediate environs of Pittsburgh lacked iron ore. George Anshultz with Colonel William Anderson had started a charcoal furnace in Pittsburgh in 1793 to cast stoves and household products, but it ceased operations in a few years. Anshultz had to ship iron ore over the mountains. Interestingly, Anshultz would marry into the Scots-Irish manufacturing clan of Pittsburgh. Pittsburgh's early foundries and nail factories brought in pig iron from Westmoreland County and other counties east of Pittsburgh. Anshultz moved to these iron ore regions of Pennsylvania and built a charcoal furnace to supply Pittsburgh. The Scot immigrants were accustomed to mining and using coal, and that experience would help change cities like Pittsburgh and Youngstown.

Many point to James O'Hara (1754-1819) as the first lord of the Pig Iron Aristocracy, and America's first "Captain of Industry." O'Hara was a Scots-Irishman with trading ties in Liverpool, from where he emigrated in the 1770s. He worked with several of the Scots-Irish fur trading companies such as Campbell's firm, using to great advantage his ties back in Liverpool. O'Hara built strong ties as an Indian trader for the Ohio Company of Virginia. He was appointed a presidential elector in 1788 and cast his vote for George Washington. From 1792 to 1794, O'Hara served as quartermaster general; he was responsible for supplying the Army of Anthony Wayne and General St. Clair's western campaign against the Indians. As a key officer in that Indian war, he was a signer of the Treaty of Greenville in 1795. O'Hara set up and built a trading network of ships and wagons to bring salt from New York to Pittsburgh. He started a distillery business in the 1790s. He is also considered the founder of the influential First Presbyterian Church in Pittsburgh. O'Hara's 1800 land holdings cover most of today's downtown Pittsburgh and the north side. In 1800, O'Hara was one of America's wealthy men with land holdings and manufacturing in several states. He was a key powerful political broker in the then "West." Politically he was a Federalist and supporter of Washington, and during the Whiskey Rebellion, he was called on to lead the militia against the whiskey rebels.

O'Hara invested in shipbuilding, which may be considered Pittsburgh's oldest industry. During the Revolutionary War, Pittsburgh was designated a government boatyard. O'Hara expanded into keelboats in the 1790s. Keelboats moved massive amounts of trade between Pittsburgh, St. Louis, Louisville, and New Orleans. He built a trading empire moving goods between Pittsburgh, New Orleans, and English ports. He even built sea-going ships in his Pittsburgh shipyards. Boatbuilding required many blacksmiths to support the operations, which helped give the area its manufacturing start. Boatbuilding also required thousands of buckets of nails. By 1802, iron products accounted for $56,548 of Pittsburgh's total value of products, and was the largest industry accounting for 16% of the total value. By 1810, iron product production had doubled and accounted for 24% of Pittsburgh's product production. O'Hara profited from the trade and built one of Pittsburgh's first mansions. He developed housing and real estate throughout the area as well. For his time, O'Hara was more dominant than Andrew Carnegie would be many decades later. O'Hara used his many ties and networks to build Pittsburgh into an industrial center. His granddaughter, Mary Croghan, married a Captain Schenley and left Pittsburgh for England, but her real estate holdings remained. At the time of her death in 1903, these real estate holdings were estimated at $50 million. It took years to sell, so as not to depress the market. Land donations include today's Schenley Park, the land where the Carnegie Library stands, and the Fort Pitt blockhouse at the Point.

In 1784 deputy quartermaster-general Isaac Craig worked for O'Hara and became an early partner. Craig was born in Ireland and immigrated to Philadelphia as a carpenter in 1741. In 1775 he joined the American Marines. He was severely wounded at the Battle of Brandywine in 1777. After the war, he became the quartermaster at Fort Pitt and married the daughter of the area's most prominent resident, General John Neville. Isaac Craig partnered with James O'Hara to build the first glassworks west of the Alleghenies in 1795, which is considered the beginning of the great industrial empire of Pittsburgh. In 1795 he reported: "Today we made the first bottle at a cost of $30,000!" Pittsburgh's rich coal seams made it a natural for glassworks, and O'Hara was one of the first to exploit it. Coal is a much better fuel than wood, which was used by most eastern glassworks. O'Hara appears to have supplied most of the capital for the business, and may have been the area's wealthiest citizen, with numerous slaves and a mansion house, although Craig was given the land by the Penn family. By 1802 the establishment was producing glass bottles, window glass, and decanters. Craig, however, would move into the background as O'Hara built a manufacturing base. O'Hara's enterprises and investments were truly amazing. Craig probably did introduce O'Hara to the pig iron business. O'Hara did have one excursion directly into the pig iron business. He partnered with John Hopkins of Vermont (a future bishop of the Episcopal Church) in taking over a failed furnace in Westmoreland County. The operation failed and O'Hara bought out his partner Hopkins to save him from bankruptcy. O'Hara also set the ideal of the industrialist giving back to the community. He was the first area philanthropist, and donated to church building funds of all dominations.

As deputy quartermaster general under General O'Hara, Craig was actively involved in the 1790s in supplying the U.S. Army of General Anthony Wayne. General Wayne mustered and supplied his army outside of Pittsburgh for his march on the Northwest Indians. Isaac Craig was working with the charcoal iron furnaces of Fayette and Westmoreland counties to supply shot and cannon for Wayne's army. Military officers such as William Turnbull and General St. Clair had built a number of these furnaces. As noted, Pittsburgh lacked iron ore to maintain operation of its early iron furnaces. While Pittsburgh was short on iron ore for smelting, it had other assets. Pittsburgh was rich in coal, which was plentiful and cheap. Coal could be used to reheat and melt pig iron for foundry operations. This type of re-melting of pig iron was a more controllable method for foundry operations versus working directly from a charcoal-smelting furnace. Re-melt air furnaces could supply more molten metal than a single tap of a charcoal furnace, especially for large castings. Re-melting also supplied a cleaner and higher quality iron than that direct from charcoal furnaces. Re-melted pig iron was also a raw material for the rolling mills of Pittsburgh. Re-melted iron was the preferred raw material for cannon production because of its quality. Shot, however, could be made more cheaply in direct cast at the charcoal furnace. The navigable Monongahela River offered the transportation network to link southwestern Pennsylvania pig iron to Pittsburgh.

Isaac Craig was a founder and charter member of the influential Masonic Lodge 45, which brought Pittsburgh's military, Scots-Irish, and manufacturers into partnership in 1785. Another Scots-Irish connection of Craig's was Isaac Meason, the iron baron of southwestern Pennsylvania's Fayette County. Meason had purchased the plantation of the famous Indian scout Christopher

Gist at Mount Braddock, Pennsylvania. In 1791, Colonel Meason improved the iron furnace to an industrial complex. Meason would be the first of the old plantation owners to move into the Pig Iron Aristocracy. Meason and his plantation was described as, "Seated as a feudal lord with complete control over Union and Mount Vernon furnaces, two forges, a grist and saw mill, two smith shops, a shoe and harness shop, and many acres of the old Gist Plantation." Meason was supplying products to markets as far as New Orleans via the Monongahela-Ohio River system. It was here in 1803 that Meason used puddling to make wrought iron out of pig iron. This successful technology was soon brought to coal-rich Pittsburgh. Isaac Meason is probably better known in Pittsburgh for his duel on Penn Avenue with future senator and Supreme Court Judge Henry Baldwin. Fortunately for Pittsburgh, both survived.

Isaac Craig's military work bought into the network of Fayette County pig-ironmasters. Alliance Furnace of Turnbull and Marmie Company in Fayette County offered a full range of operations including a charcoal furnace, foundry, and forge. Owner John Turnbull had an active mercantile business and had formed a whiskey production partnership with Isaac Craig in the 1788s. Turnbull had sawmills and salt works in the area as well. The Alliance Furnace was a major factory capable of producing castings, forgings, and blacksmithing. Eastern Pennsylvanian financier Robert Morris was an investor in the operation. Furthermore, Turnbull and Marmie had a distribution network tying Pittsburgh and Philadelphia. Isaac Craig and Turnbull teamed up to be a major supplier to the army of Anthony Wayne. Craig, as well as James O'Hara, clearly developed their ties and businesses through their government quartermaster work. O'Hara seemed to have the advantage of capital, and Isaac Craig seems to have run into financial problems in the recession of the early 1800s. O'Hara appears to have came to America with some financial wealth. In the early 1800s, England declared a type of economic war, dumping iron and glass products, with the goal of destroying America's infant industries. The result was extensive bankruptcies of western Pennsylvania, eastern Ohio, and northern Maryland prior to 1810.

Another prominent Scots-Irish businessman was John Wilkins. The John Wilkins family had come to Pittsburgh in the 1780s and was an early Pittsburgh investor. Wilkins, like O'Hara and Craig, served as a quartermaster general in the United States Army. John Wilkins Jr. opened a window glass factory in the early 1800s, and was a director with O'Hara in the formation of the Bank of Pennsylvania in 1802. John Wilkins's younger brother founded the Bank of Pittsburgh in 1810. John Wilkins Sr., along with the O'Haras, Dennys, Craigs, Bayards, and Brackenridges, are considered the founding families of Pittsburgh Presbyterianism. Wilkins Jr. would serve as quartermaster general from 1796 to 1802, strengthening his family ties with the O'Hara trading empire. In addition, the Wilkinses, O'Haras, Dennys, and Brackenridges were considered the founding families of the Federalist Party in Pittsburgh. William Wilkins, the younger brother of John Jr. would become a key politician in the Pig Iron Aristocracy. John and William were Jacksonian Democrats but were strong protectionists until the 1840s. William would serve as a Senator, U.S. Representative, Supreme Court Judge, Ambassador to Russia, and Secretary of War. John Wilkins's sister, Nancy, married another Pittsburgh founder and Scots-Irishman, Ebenezer Denny. Pig iron politicians were often given cabinet positions to pander to the iron districts in Pennsylvania and Ohio. This mixing of Federalists and Jeffersonian Democrats, like Germans and Scots-Irish, was common, but the protectionist beliefs were never compromised by the pig iron regional politicians.

Ebenezer Denny was O'Hara's real estate partner in the 1790s. Denny's ties had gone back to supplying the British at Fort Pitt. Although he was English, Denny was a member of the Presbyterian Church. Denny served with O'Hara and others as a director of the Bank of Pennsylvania, a founder of the First Presbyterian Church, and a member of Masonic Lodge 45. He was also a veteran of the Revolutionary War. In 1816, Denny became the first mayor of Pittsburgh. The Denny family remained a prominent Pittsburgh family for decades and was a cornerstone of the Pig Iron Aristocracy. Denny's son, Harmar Denny, would marry Elizabeth, the daughter of James O'Hara. Harmar was invested in many of the O'Hara's companies and had also served as a quartermaster in the army. Harmar Denny would help found the Pittsburgh and Steubenville Railroad and the Pennsylvania Canal as well as the Western University of Pennsylvania (a forerunner of the University of Pittsburgh). He founded the socially elite Chemical Society, which fostered chemistry, mineralogy, and the physical sciences. The Denny and Wilkins families were typical of the split in western Pennsylvania politics that took some industrialists out of the Federalist column. Denny even helped with the formation of a church for poor Catholics, who numbered less than 300 in 1825. It took the Whig Party in the 1840s to unite the industrialists under a protectionist banner.

O'Hara, on the other hand, expanded his interests into sawmills, tanneries, salt works, shipbuilding, and gristmills during the same period. In 1803 O'Hara built Pittsburgh Point Brewery, which made a type of porter beer popular with the Scots-Irish. He teamed up with Ebenezer Denny to build row houses for the city, and by the time of his death, O'Hara was the largest real estate owner in the area. O'Hara also moved into finance and banking. In 1810, pig iron shortages rose as President Jefferson placed an embargo on British products. The decimated iron furnaces struggled to re-open to meet domestic demand. O'Hara bought Hermitage Furnace from General Arthur St. Clair, who had defaulted on a bond from O'Hara. The upcoming War of 1812 created a pig boom bringing on furnaces in Pennsylvania and Ohio. Many Scots-Irish iron families moved up from Maryland and Virginia to start furnaces in western Pennsylvania and eastern Ohio. One of these families was the ancestor of future President William McKinley. In 1805 another pig iron lord, Joseph McClurg, opened Pittsburgh's first foundry. This was Pittsburgh Foundry, located at Smithfield Street and Fifth Avenue. McClurg became one of the premier suppliers for the American Army and Navy of 1812. In 1810 McClurg purchased a number of bankrupt Westmoreland furnaces to supply pig iron to his Pittsburgh foundry. This combining of Pittsburgh banks, foundries, and rolling mills with the southwestern

Pennsylvania mountain furnaces became the infrastructure of the area's Pig Iron Aristocracy.

Joseph McClurg had come to Pittsburgh with his Scots-Irish family in the late 1790s. The family settled on Pittsburgh's south side. He had some theological training in Reformed Presbyterianism. The McClurg foundry had two air furnaces to remelt southwestern Pennsylvania pig iron. It appeared to have been financed by the new branch of the Bank of Pennsylvania, which opened in 1805. The bank branch had been established with James O'Hara as president and Ebenezer Denny, Joseph Barker, Anthony Beelen, Thomas Baird, Boyce Irwin, David Evans, and George Wallace as trustees. O'Hara was clearly a visionary, seeing that many of the products brought in from the east could be manufactured in Pittsburgh. Shovels, axes, hatchets, stoves, andirons, spades, and particularly nails. O'Hara's bank not only financed McClurg's foundry, but two nail and slitting mills. By 1810 there were six naileries having a product value of $49,890 or 33% of Pittsburgh's metal industries. Nail tonnage was about 200 tons per year.

Wire was another product in high demand on America's growing frontier. German-born Peter William Eichbaum (William Sr.) had been brought to Pittsburgh in 1797 to manage James O'Hara's bottle glass factory. Eichbaum turned out to be an extremely talented engineer. He mastered bottle making, but moved into other industries and built a wire mill in 1810. Lacking full financing from O'Hara, he petitioned the state legislature, which was interested in developing manufacturing. The legislative action stated:

> Whereas it is true policy of this State to give encouragement to works of public utility, to foster our own manufacture and to render ourselves independent of foreign nations for articles of absolute necessity; therefore, be it enacted. That a loan be granted from the State to William Eichbaum of three thousand dollars.

The use of a state loan was typical of the time, and James O'Hara used this route as well to expand the iron industry of the area. Eichbaum's wire mill was on the Monongahela River beside O'Hara's brewery. It was a state-of-the-art mill, using steam power to drive the rollers. Steam engines were being built en mass in Pittsburgh in 1807 at the plant of Oliver Evans, Mark Stackhouse and Mahlon Rogers. The company was known as Pittsburgh Steam Engine Company. Eichbaum a few years later invented a sheet rolling mill that produced world-class sheet metal. Water power and later steam engines inspired more Pittsburgh rolling mills prior to 1812. Steam power allowed for concentration of manufacturing in a centralized area. Prior to steam, the power of streams and rivers was needed. Countryside iron processing operations could be brought to the central city with steam power. Scots-Irish Christopher Cowan started a sheet mill near Pittsburgh in Shadyside in 1806, and Jeffery Scaife opened a tin plate mill on Pittsburgh's Market Street around 1802. The Scaife family would become one of Pittsburgh's oldest and richest. In 1927 the Scaife family and Mellon would combine in marriage and create one of America's greatest fortunes in Pittsburgh.

Pennsylvania's state legislature had been taken over by industrial interests by 1800. Pennsylvania was the major pig iron producer with furnaces across the state. There were iron furnaces in the east, central, and southwestern regions of the state. The Jefferson trade embargo of 1807 created a shortage and boom in pig iron production. With the Pig Iron Aristocrats' control in the state legislature and the old military federal ties, Pennsylvania pig iron production expanded as did Pittsburgh pig iron users.

Pittsburgh became known as "Iron City," not because of its iron production, but from its use of iron. Whether it was wire, cast steam engine parts, or nails, pig iron was needed. The real boom in the use of pig iron came with the opening of a puddling furnace in 1819 by Union Rolling Mill. Englishman Henry Cort had invented puddling in 1784, but it took another twenty years for it to evolve into an efficient method. Puddling furnaces heated pig iron to a gooey mass, which was worked as a ball to remove carbon and produce wrought iron. A skilled puddler would use a hooked rod to pull a pasty mass out of the furnace. The pasty ball was moved to a forging hammer that squeezed and hammered out slag while burning out carbon. A puddling furnace would melt a five-hundred-pound "heat" of pig iron. A puddler could process four to five heats into wrought iron in a nine-hour day. The amount of wrought iron produced exceeded the fifty- to three-hundred-pound wrought iron output of heat and beat process forges. This puddled wrought iron was the raw material for rolling and nail mills. Puddlers were the top paid craftsmen in the country, making as much as twenty-five dollars a day, compared to common laborers, who made $1.25 a day. Puddling made wrought iron production many times more productive. The Union Rolling Mill would become America's largest consumer of pig iron until the arrival of Carnegie's steel mills in the 1870s. The puddling process would morph into steel mills of the 1860s.

Besides puddling furnaces, seven wrought iron rolling mills were built from 1819 to 1825. Besides Union, there was Grant Hill Iron Works, powered by an eighty-horsepower steam engine made locally at Columbian Steam Engine Company. Grant Hill manufactured a variety of wrought iron bar. Sligo Rolling was built by Robert Stewart, who was from an iron plantation family in the Juniata region of Pennsylvania. Sligo made musket barrel bars for Harper's Ferry Armory. On Pittsburgh's Pine Creek, there was a water-powered rolling mill owned by M. B. Belknap. The Juniata Rolling Works used a 120-horsepower engine built by Matthew Smith of Pittsburgh's Benny and Smith.

Iron nails were the tonnage product of Pittsburgh manufacturers in the 1820s and 1830s. The switch from log cabins to frame housing had caused a super boom in nail demand. Actually, it was more of a co-dependency, as volume nail production made frame building economical. Prior to 1800, blacksmiths made nails by a painstakingly long process. As volume increased, blacksmith shops grew into small factories, and then into naileries. Thomas Jefferson started his own nail operation using slaves at Monticello. Nails made Pittsburgh the "Iron City" because of some natural and manmade strategic advantages. Pittsburgh had three nail factories by 1806, but most of these were extended blacksmith operations of southern Scots-Irish immigrants. These Scots-Irish ironworkers were also moving into eastern Ohio valleys, following new ore deposits. The resultant pig iron was sent via family ties to Pittsburgh for processing.

Most important was the invention of a mechanical nail making machine by Jacob Perkins of Massachusetts in 1795, but its application took a decade. The Perkins machine could produce up to 500 nails per minute. This Boston technology was brought to Pittsburgh by two Boston ironmakers: William Stackpole and Ruggles Whiting. In 1811, they built a nail mill around 1812 known as Stackpole and Whiting. Pittsburgh offered a good supply of pig iron and steam engines. British dumping of iron products in the period after the War of 1812 put Stackpole and Ruggles in bankruptcy. Amazingly Stackpole and Ruggles was considered the most modern nail mill in the world. By 1825, there were additionally one hundred or more patents improving the nail making machine (many of them from Pittsburgh). Pittsburgh was strategically connected to the furnace operations of the nearby mountains and Ohio's Mahoning Valley. Key also was the abundance of coal in Pittsburgh for fuel in puddling furnaces for wrought iron production. Coal also fired the locally made steam engines that drove the iron rolling mills. Amazingly, there were eight steam engine companies in Pittsburgh by 1830, and it led America in engine manufacture. Lastly, Pittsburgh's river connection to the west tied the market demand to the operations. Pittsburgh stood almost alone in the unique combination of resources and technology, but the pig iron economic network ran through Ohio, Maryland, Virginia, Kentucky, and Pennsylvania. King Pig Iron arrived long before King Cotton.

In 1824 Peter Shoenberger built Pittsburgh's successful volume nail mill known as Juniata Works. Juniata Iron was on the Allegheny River at Fifteenth Street near the terminus of the Pennsylvania Canal, in which Shoenberger had a financial interest. The Pennsylvania Canal connected Shoenberger's pig iron furnaces in the Juniata district of southwestern Pennsylvania to his rolling mill. True to his iron plantation heritage, he built a large mansion near the plant. By 1832, "Pittsburgh Peter" Shoenberger built a nail operation about fifty miles downriver at Wheeling, West Virginia, one of the other strategic locations. Mahoning Valley in Ohio also had the right combinations for nail production. Shoenberger was the son of a Pennsylvania iron plantation owner in the Juniata region. Shoenberger had built a large forge and bloomer with his charcoal furnaces. His Juniata Works in Pittsburgh used a nail machine built by iron engineer John McElroy. The works had six nail machines and a spike machine, newly invented by McElroy. The works employed eighty men and produced six tons of nails a week.

Peter Shoenberger came from one of Pennsylvania's German ironworking families. Originally the family worked as blacksmiths and ironmongers in Lancaster. Coming over the mountains, Shoenberger's father started one of the most successful iron plantations in Pennsylvania. The Shoenbergers would own pig iron furnaces through the tri-state ironmaking region of Ohio, Pennsylvania, and West Virginia. Peter Shoenberger would be the German "O'Hara" of the 1830s in Pittsburgh; his manufacturing empire even surpassed that of O'Hara. In 1831, Shoenberger formed Shoenberger and Company with his son, John Shoenberger, to produce iron and steel products. In the 1830s, both Peter and John invested in stagecoaches and canals. By 1833, the firm was pioneering the making of blister and crucible steel, and would father some of the area great steelmakers. The Shoenbergers also were the first in the area to move into wire production, which would lead to the formation of American Steel and Wire. Peter and John would foreshadow Andrew Carnegie in the use of vertical integration, that is, owning the whole supply chain, such as coalmines, coke ovens, iron furnaces, rolling mills, and the transportation system. John Shoenberger became a director of the newly established Merchants' and Manufacturers' Bank. The Shoenberger family would also be part of the iron industry expansion into Wheeling, West Virginia, and Youngstown, Ohio. Peter Shoenberger's daughter would become the wife of Pittsburgh songwriter Stephen Foster. Peter Shoenberger was a founder both Lutheran and Episcopal churches as well as the social Pittsburgh Club. He had a pew in the First

English Lutheran Church. John Shoenberger was involved with a number of banks, factories, steel mills, hospitals, and churches. He would be the first of Pittsburgh's great art collectors, and an amateur astronomer (like James O'Hara). Shoenberger was a founder of the Allegheny Observatory. The Shoenbergers would operate iron and steel operations into the 1900s.

CHAPTER FIVE

The Political Rise of the Pig Iron Aristocracy

The rise of the Pig Iron Aristocracy and the pig iron industry was directly related to the political foundation of Henry Clay's "American System." Clay's American System was a protectionist approach to shield American industries via tariffs and an aggressive approach to national improvements. Clay would become the Zeus of the iron industry gods. To fully grasp the roots of Clay's economic philosophy, one must first understand Thomas Jefferson's vision of the nation. Jefferson envisioned an agrarian society of farmers and merchants. In 1790, an estimated 90% of the American population was employed in agriculture. Jefferson's vision demanded free trade to assure that American crops could move into foreign markets readily. He had grown up in a tobacco and cotton culture that depended on European purchases. While he believed in farm self-sufficiency, he feared the industrialization that he had seen in Europe. Hamilton, on the other hand, saw America's freedom rooted in its ability to achieve economic freedom through manufacturing and banking. Hamilton, the soldier, was well aware of the role of technology and manufacturing in the ability of a nation to win wars, and believed that manufacturing was fundamental to America's freedom. Hamilton as a young officer found the colonial army constrained by lack of iron cannon and rifles because of the lack of American manufacture. Interestingly, President William McKinley would have the same experience in his regiment during the Civil War, and it would be during President

McKinley's administration that the golden era for pig iron protectionism flourished. Hamilton, however, remained a free trader like many of his Federalist friends, believing that financial systems were the basis for industrialization. Actually, the Federalists were split on tariffs, some seeing no need for them because America lacked manufacturing, and others seeing tariffs as a source of federal revenue.

Both Jefferson and Hamilton were constrained by the agricultural nature and lack of manufacturing in America at the time, as well as by earlier colonial British constraints on industries such as ironmaking. These British prohibitions such as the British Iron Act of 1750 had infuriated Scots-Irish ironmakers like James McKinley (William McKinley's grandfather) in western Pennsylvania. In particular, the Scots-Irish moved to the Ohio frontier to avoid tax laws. The Iron Act of 1750 allowed for all raw bars of smelted iron known as pig iron to be shipped to England duty-free, but outlawed the production of iron products, such as kettles, skillets, stoves, forged iron for guns, and steel for the blacksmith shop. These frontiersmen remembered and vowed never to be economically restrained again by any government. Many of these same Scots-Irish would flee western Pennsylvania to Ohio, Kentucky, and Tennessee to avoid the federal tax on whiskey manufacture in 1794, and would become part of the political base of frontier politician Henry Clay. It would be a base that often disagreed with Clay, wanting cheap imported goods.

The Whiskey Rebellion more than anything caused a political divide in the Scots-Irish along economic lines. The wealthy Scots-Irish industrialists and the Presbyterian Church leaders had supported the Federalist application of the law. Their poorer cousins in the Pennsylvania hills opposed a strong central government and taxes. The whiskey tax forced these frontier Scots-Irish into Jefferson's "Republican" Party. In the 1820s, "the clapboard junto" of Pittsburgh (those who lived in clapboard houses) put together a strong opposition to the Federalist manufacturers. The Jacksonian Democratic Party (known as the clapboard democracy) controlled the area, but Pittsburgh's congressmen were protectionists, who sided with Henry Clay in Congress. Pennsylvanian Henry Baldwin supported Henry Clay's American System with an iron fist. Pennsylvania Senator Judge Wilkins became known as the "iron knight" in his support of iron tariffs. The manufacturers ultimately wrestled the vote away from the Democrats as Andrew Jackson's policies turned anti-manufacturing. These manufacturers often held dinners for national protectionists, such as Mathew Carey and Henry Clay. Nearby manufacturing cities such as Steubenville, Ohio, and Wheeling, Virginia, developed a manufacturing network through Clay and Carey.

If President William McKinley's economic roots can be traced to Henry Clay then Clay's roots can be traced to Federalist and first Secretary of the Treasury, Alexander Hamilton. And like McKinley, Hamilton learned his economics as an army supply officer. Hamilton, while on Washington's staff, had struggled to get clothing for the Continental Army soldiers because of America's dependence on British goods. He also learned the hard lesson of inflated dollars as merchants rejected government notes. His experiences would be the foundation for Hamilton's classic in 1791, the *Report on Manufactures*, which "prophesied much of post-Civil War America."[8] It would augur both Henry Clay and McKinley's approach to government as it related to national industrial planning, and the importance of protecting such industries as pig iron. Hamilton was the first to suggest a scientific approach to tariffs versus across-the-board revenue tariffs. First defense and national industries were to be protected, followed by targeted infant industries. He argued for lower tariffs on raw materials to help industry. Hamilton would win many disciples including Henry Clay.

Henry Clay was a Virginia lawyer who moved to Kentucky to launch a career. In 1810, he was elected to the United States Congress. Clay was a nationalist, Patriot, Republican, and Federalist. In his junior years in the Senate, he advocated a strong national bank and a national road system. Often Clay favored the

---

[8] Ron Chernow, *Alexander Hamilton* (New York: Penguin Books, 2005), 374.

good of the nation over his own constituents. His oratory, compromising skills, and patriotism brought him quickly to the position of Speaker of the House. Clay not only fashioned the position of House Speaker, but he formed the powerful standing committees such as the Ways and Means Committee, which would be the pedestal to launch the career of William McKinley years later. Clay also created a Committee on Manufactures to help stimulate American manufacturing. Clay appointed members for these powerful committees, and thus centralized legislative power under the position of Speaker. Clay used the power to create a national infrastructure for an industrial America. Clay's vision of an industrial empire took him from Jeffersonian Republicanism to Federalism and then to conservatism. Federalists, however, were New England based free traders. Clay's arguments and the rise of American manufacturing won over many Federalists who believed in the destiny of the American republic as a world power.

The struggle and the delineation of these competing visions of agriculture and industry would evolve as the nation evolved. By the dawn of the 1800s, the nation had a developing manufacturing sector in New England. America was learning to produce guns, gunpowder, farming implements, and textiles. Even Jefferson marveled at the industrial growth and its contributions to the nation. Yankees had smuggled in new automated looms from England and American textile manufacture moved to a new level. The acceptance of automation had given American textile manufacturers an advantage over the labor-intensive British industry. Its own anti-automation proponents known as Luddites had held England back. Furthermore, the War of 1812 had caused a surge in American textile production as part of the need for economic freedom, as well as a boom in iron manufacture in the middle states. The McKinley family would ultimately purchase one of those infant iron furnaces of 1812 in Niles, Ohio, in the 1850s.

The War of 1812 and the economic warfare that followed extended into the 1820s, and proved Hamilton's view of the need for economic independence. The British attempted to destroy the American textile industry by dumping huge quantities of British textiles on American docks. The British similarly dumped cheap pig iron to suppress the American pig iron industry. The British were more successful in this type of war, bankrupting hundreds of American manufacturers and closing charcoal iron furnaces throughout the country. Lord Brougham in Parliament of 1816 summarized the strategy: "It is well worth while to incur a loss upon first exportation, in order by the glut to stifle in their cradle those rising manufacturers in the United States which the war has forced into existence contrary to the natural course of things." The Northeast put political pressure on Congress to save its textile-manufacturing base, and the Pennsylvania, Virginia, and Ohio pig iron producers and users joined the political pressure.

Congress hesitated to act, torn by competing regional goals. The Southern cotton growers opposed any tariffs on British goods, believing Britain would retaliate with tariffs on cotton and tobacco. The major portion of the South's cotton and tobacco went to Great Britain for processing. Furthermore, even the Northeast representatives were torn between the textile manufacturers and the merchants, who favored free trade. The struggle in Congress in 1816 would produce a new champion in Henry Clay, Speaker of the House. The Congress appeared hopelessly deadlocked on the issue. Clay built an alliance for the tariffs based on nationalism versus regional politics. The debate took place in a temporary brick building (at the site of today's Supreme Court), known as the "Old Brick Capitol." The city of Washington lay in ruins after the sacking by the British, and offered a stark reminder to Congress of the need for a strong defense. Clay found allies in Southerners John Calhoun and President Madison, who would help tip the balance. Clay brought in the middle state representatives who had suffered from British dumping of iron products to suppress American industry. He astutely played on rising nationalism and anti-British sentiments to bring in enough Southern votes to pass the tariff. The embryonic Pig Iron Aristocracy rallied behind Henry Clay. The result was America's first tariff: the Tariff of 1816, which established duties of 25% on cotton and wool products and 30% on iron products. The Tariff of 1816 would ensure the financial security of the

Pennsylvanian ironworker family under the future President William McKinley.

With the Tariff of 1816, Clay inaugurated the "American System" of focused, protective, and selective tariffs for the good of the nation. Henry Clay told Congress in 1820:

> In passing along the highway one frequently sees large and spacious buildings, with the glass broken out the windows, the shutters hanging in ruinous disorder, without any appearance of activity and enveloped in solitary gloom. Upon inquiry what they are, you are almost always informed that they were some cotton or other factory, which their proprietors could no longer keep in motion against overwhelming pressure of foreign competition.[9]

It was a story that could have been written today, and one that Clay wanted to change.

Clay had not only broken with Jefferson's thinking, but that of his own Federalist leanings of free trade. The Federalists were split because tariffs appeared to be a heresy. Federalists were free-trader Yankees, even though they favored helping national industries. The split would eventually lead to the Whig Party. Clay forged a new path for American capitalism that was nationalistic and economic. It was a Magna Charta of American economic freedom. Clay realized that economic war was a reality in the world of the 1800s. Clay's vision was similar to Jefferson's, differing only in that industry was substituted for agriculture. Like Eisenhower in the 1950s, Clay envisioned a system of national transportation to support industrial growth. Clay molded a powerful new philosophy, which blended Jeffersonian independence with economic manifest destiny. Clay went further to justify his American System by blending in American moral superiority with nationalistic capitalism. The momentum had turned in Clay's favor by 1824. The struggle, however, would not end, as skillful opponents such as Daniel Webster arose.

---

[9] Thomas Cochran and William Miller, *The Age of Enterprise* (New York: Harper & Row, 1942), 12.

Congress in 1824 moved to debate even more extensive tariffs. Clay, the orator, would emerge as leader of this industrial movement. He thundered in Congress with an oratory reminiscent of Patrick Henry a generation earlier:

> Is there no remedy within the reach of the government? Are we doomed to behold our industry languish and decay yet more and more? But there is a remedy, and the remedy consists in modifying our foreign policy, and adopting a genuine American System. We must naturalize the arts in our country, and we must naturalize them by the only means, which the wisdom of nations has yet discovered to be effectual- by adequate protection against the otherwise overwhelming influence of foreigners. This can only be accomplished by the establishment of a tariff.[10]

Clay struggled against his oratorical match, Daniel Webster, and the entire Southern wing of the House of Representatives. Clay had strong support of the Pig Iron Aristocracy's Henry Baldwin, the "iron knight," from Pittsburgh. Daniel Webster of New Hampshire opposed the tariff because it might hurt the New England shipping industry. Clay persisted, and on April 16, 1824, the Tariff of 1824 passed 107 to 102. Since cotton and tobacco caused resistance in South, a breakdown of the slave versus non-slave states is more telling. All non-slave states: for tariff 89; against 32. Slave states: for tariff 18; against tariff 70. Ohio supported the tariff not only in the Mahoning Valley but also in more western counties where an infant wool industry was emerging. President Monroe signed it in May. The Tariff of 1824 extended the general level of protection to 35% ad valorem (the percentage of value as represented by the invoice). The tariff included cotton, wool, and iron products, also extending it to the hemp producers of Clay's Kentucky.

Clay's politics started a change in what would become the future core of McKinley's base: the Mahoning Valley, Niles, Youngstown, Canton, Pittsburgh, western Virginia, and Ohio's Western Reserve. These old frontier areas had a large Scots-Irish population who had opposed with guns the whiskey taxes of the

---

[10] Robert Remini, *Henry Clay: Statesman of the Union* (New York: W. W. Norton & Co., 1991), 230.

Federalists. They tended to be frontier Jeffersonians, but the protection on wool and iron by Clay started to build a base for a new type of Federalist. Industrialization was changing the area as well, and most middle-state Federalists were moving toward protectionism versus their initial free trade policy. National roads and canals favored the growth of these areas, which was fundamental to Federalist theory. This Ohio old frontier was also similar to Clay's Kentucky congressional district.

The victory of the 1824 tariff bill split the nation as well and cost Henry Clay the presidency in 1824. Clay would further develop his American System as Secretary of State for the new president, John Q. Adams, but political opposition in the South was growing too. The opposition was gathering behind Andrew Jackson and the Democrats. While the Pig Iron Aristocracy supported Adams and Clay, Jackson's popularity triumphed over local interests on the Allegheny plateau of Pennsylvania and Ohio. In fairness, at the time Northern state congressmen, whether Jacksonians, Jeffersonian, or Federalist, supported tariffs. The rural Scots-Irish were fierce Jacksonians as a results of Federalists efforts to tax whiskey. The Jacksonians were positioning for a presidential run, and found success early by taking control of the Twentieth Congress in 1827. The Jacksonians of the Democratic Party had strength in the west and south. The result of the popularity of tariffs in the east and middle states and political division led to the "Tariff of Abominations" as tariffs became a party and political issue. The Democrats actually allowed an unbalanced tariff to be passed by the National Republicans, turning the measure into future votes for the Democrats and ultimately lower tariffs. The bill extended the tariff on certain products from Ohio, Pennsylvania, Kentucky, and New York by increasing the duties on iron, spirits, hemp, and molasses. The wool products of New England, which Clay had initially used to justify the earlier tariffs, were basically ignored with a modest increase. Clay could only watch as the bill passed, assuring future Jackson votes in the powerful northeast. President Adams signed it because its usefulness outweighed future political problems. The Tariff of Abominations would in retrospect give Jackson and the free trading Democrats the White House, set back the nationalistic tariff policy of Clay, and move the nation towards civil war. The Jacksonian movement re-strengthened the Democrats on the Ohio frontier because of the personal popularity of Andrew Jackson, but ultimately Jackson proved a weak supporter of American manufacturers, which helped bring some Democrats into the Whig Party of Henry Clay.

To focus solely on the political ramifications of Clay's American System, however, would also be to overlook the realization of Clay's (and ultimately McKinley's) dream of an Industrial Eden. And it truly was a system where tariffs were focused to help infant industries, and the tariff revenues were used to build roads and canals. In the Northeast, textile mills were growing; in Pennsylvania and Ohio iron furnaces were being built; and the American nation was moving from an underdeveloped country to an industrialized one in the first decade of the 1800s. The American System of industrialization was being held by industrial critics such as Charles Dickens as Utopian. The manufacturing methods and automation of American industry was rapidly becoming the standard of efficiency for the world. Pioneering American industrialists such as Francis Cabot Lowell started to develop uniquely American textile factories. While still physically demanding, the factories were clean and offered schooling and training. Even old Jeffersonians were proud of the rise of American manufacturing supremacy.

Part of the superiority could be found in the American "factory system." The tariffs and government contracts produced volume levels never previously attained, allowing a shift in many industries from a crafts system to the factory system. In 1812, Thomas Jefferson had contracted Eli Whitney to produce arms with interchangeable parts; it was the first government contract. The Springfield Armory under Colonel Roswell Lee advanced the factory system in the 1820s with the application of labor specialization and automation. The real progress was not so much in the ability to produce standardized and large quantities of weapons, but the growth of the new industry of machine making to support such industries. The Tariff of 1824 had built a foundation for American investment by stabilizing the market for American industrial goods. In the 1830s, America was becoming a nation of mechanics as exemplified by the appearance of mechanics' institutes, schools, magazines, and newspapers. The Franklin Institute of Philadelphia was founded in 1824 to promote the advance of mechanics and science.

The protective tariffs caused an industrial boom in New England's textile industry. The textile manufacturers were well protected and the textile industry grew, and it gave birth to the machine industry. Utilizing the water power of the Connecticut River, the machine industry grew to support textile manufacture, arms manufacture, and farming equipment. The growth caused England's greatest machinists to immigrate to New England. They came with British machines, which the New England Yankees reverse engineered into better machines. At the 1851 Crystal Palace Exhibition, the British were shocked and humbled by the American machine technology. The Connecticut Valley would become a bastion of tariff support for Clay and then McKinley. The machine industry opposed the Jacksonians and would ultimately find a home in the Whig Party and Republican Party.

The tariff-created volume and stabilization in the textile industry ushered in an era of invention and technology after 1812. An 1837 survey by the state legislature of Pennsylvania of the textile industry reported the amazing advance:

> Ten years ago, it was generally supposed, that few improvements in machinery could take place. The machinery of that day is now useless; and another period of ten years may make the same difference; manufacturers are subject, in this particular, to a heavy tax. He who advances with the times, must incur the cost of continual improvement; he who lags behind, must lose in the cost of his production. The success of the American textile industry was a manufacturing miracle attracting the world's manufacturers, writers, and politicians.[11]

---

[11] Anthony Wallace, *Rockdale* (Lincoln: University of Nebraska Press, 1972), 186.

Henry Clay toured the textile industry several times in the 1830s to furthermore promote his "American System." One of the mills was named after Henry Clay to honor his protective tariffs. Even Clay's enemy Andrew Jackson honored the textile industry with a personal tour. Clay was now able to address the hero of free trade and the Democrats, Adam Smith, a Scot. Adam Smith's 1776 book, *The Wealth of Nations*, had become the banner for free trade. Henry Clay now argued that free trade could reduce prices in the short run but at the expense of capital investment, invention, and automation. Furthermore, Clay saw capitalism as a national philosophy, not an attribute of free trade. Still, the farming majority saw it much differently, fearing international reprisals and higher prices for domestic goods.

Henry Clay would bring philosophical support for protectionism. Henry Clay, the lawyer, had actually read Smith's long and boring narrative. Clay understood the weakness of Smith's ideal of free trade. Smith had allowed for a number of exceptions in his free trade proposals such as defense industries. By doing so, Smith allowed exceptions for some of Britain's strongest industries. Clay saw industry itself as basic to America's defense of its freedom. Clay believed in capitalism at the national level, but felt political factors restricted the type of international capitalism suggested by Adam Smith. Furthermore, Clay was a true conservative, seeing America as having supremacy and a God directed destiny. The survival of democracy depended on the productivity of its laborers. To Clay protectionism was critical to freedom and democracy itself. He further argued that Smith based his thesis on the trade-based economy of Scotland, where trade was their industry. Clay's friend and amateur economist John Q. Adams agreed completely. Both of these men would champion the "American System" as a political alternative. Clay's philosophical base for protectionism helped win over some old Federalists as well as many Jeffersonians.

Another American manufacturing movement of the 1820s was the development of utopian manufacturing communities. These movements came from Europe as part of the reaction to the miserable working conditions in factories. These manufacturing communities were actually Jeffersonian in nature. They were self-sufficient communities, growing food and manufacturing implements. These manufacturing communities included the Shakers of Kentucky and New England, New Harmony of Indiana, Economy of Pennsylvania, Zoar of Ohio, and Oneida of New York. Henry Clay had invested in the Shaker community near his home in Kentucky. The success of these communities resulted in allowing their excess to be sold to surrounding communities as well as investing in other American manufacturing. Before its closing in the 1870s, Economy Village in Pennsylvania was a major stockholder of the Pennsylvania Railroad. Clay never tired of touring these industrial communities. He was convinced that clean, comfortable living was compatible with American manufacturing.

Clay's industrial vision was growing in 1828, but it lacked a national political base. The success of the American System was still a regional phenomenon, allowing Democrat Andrew Jackson to take the White House in 1828. The Jacksonians represented the Jeffersonian legacy of free trade and low tariffs, drawing support from the south and west. The Jacksonians' efforts to reduce tariffs were somewhat muted in Congress by Clay and his followers. Jackson's popularity won him re-election over a challenge by Henry Clay. The Tariff of Abominations in 1828 had caused the legislature of South Carolina to pass a Nullification Act. For the sake of the Union, Henry Clay compromised with the Jacksonians to pass a slightly reduced tariff. The Jackson administration was a constant problem for protectionists such as Clay, yet the industrialism of the north and east represented a growing political base. The Jacksonians would bring down the National Bank system, but tariffs remained with only minor reductions.

Clay's hypothesis would be proven over and over again as the nation's tariff rates varied up and down. Economists and historians have ignored the evidence since free trade economics came in vogue after World War II. The 1840s, 1880s and 1890s are decades that appear to be designed experiments on the efficiency of tariffs in promoting growth, invention, and jobs. In the 1840s, railroad rails accounted for one third of the demand, and the manufacture of this iron tells a vivid story of the impact of tariffs. In the 1832 Tariff, the Congress exempted railroad iron from the tariff concluding that American rails lacked the quality, technology, and cost to compete with the rails of Britain. Almost all railroad iron was imported from Britain. Britain had the rolling mill technology to produce the T or H rail, while America lacked the ability to roll these shapes. Britain's coal produced iron was also superior in quality because burning coal instead of wood made it impurity free. The American iron industry was driven by low-tech heating using hardwood because of its availability, while Britain had adopted coal gaining quality and efficiency. In the 1840s, the American iron industry was consuming more forest acreage than Brazil does today. Interestingly, Brazil still uses wood to make iron because of the availability of rain forest wood.

Henry Clay was one of the few in Congress that had not given up on the capability of America to produce railroad iron. The Whigs and Clay took over both houses of Congress in 1840, setting up the Tariff of 1842, which specifically included railroad iron as well as restoring the overall tariff levels of 1832. It would be the last tariff bill of Clay, who resigned prior to its actual passage. As a result of the high tariff, the investment in iron rolling technology and coal furnace technology boomed. By 1844 American mills such as Mt. Savage in Maryland were rolling the first iron T rails. Furthermore, that year a number of iron furnaces such as Phoenix Company were making coal smelted pig iron.[12] In 1847, Congress reduced the tariff, allowing Britain to flood the rail market with cheap iron to keep their furnaces running. By 1850, of the fifteen rail companies that had started in America due to the Tariff of 1842 only two were still in production. The early 1850s required Britain to take iron production away from America to support the needs of the Crimean War. American market prices rose due to a shortage, bringing American producers back into the market. The experience gave rise

---

[12] Thomas Cochran, *Frontiers of Change* (New York: Oxford University Press, 1981), 106.

to Matthew Carey, who would win a young Abe Lincoln over to protectionism.

Thanks to the Tariff of 1842 and its reinforcement in the Lincoln administration, an American iron rail industry was developed in time for the railroad boom of the 1870s, which saw the rise of Carnegie Steel. Until the 1870s, British iron rails ruled because of their price and quality. The 1842 tariff allowed Americans to make the necessary investments and develop the technology to improve quality. Clay realized that without protection or some form of subsidy, strategic American industries would never develop. In the 1870s and 1880s, the continued Republican tariffs allowed the infant American steel industry to blossom with the world's largest and most technologically advanced rail mills at Johnston and Braddock, Pennsylvania. By 1898 American steel ruled the world and employed tens of thousands of workers, and American steel rails were being shipped to China and Europe. During this period of protectionism, the miracle of steel technology emerged from iron production, American furnace sizes dwarfed their European counterparts, American steel quality was uncontested, and the American steel industry overtook all other industrialized nations. Henry Clay's vision for iron and manufacturing would be fully realized during the McKinley administration.

Clay's American System and his Whig Party slowly made inroads into the politics of western Pennsylvania and Ohio. Clay's run for the presidency in 1844 united the region of iron manufacturers under the Whig Party. While Clay lost nationally, he carried Pennsylvania's Allegheny County and Ohio's eastern counties. Many argued that the Pig Iron Aristocrats hired all the Democrats possible to deliver coal barges down river on Election Day in 1844. A multi-state region roughly defined, as a triangle connecting Pittsburgh, Wheeling, and Youngstown became the iron triangle and stronghold of the Whig Party. This was the heart of pig iron manufacturers and the Pig Iron Aristocracy. Now even the rural Jacksonians joined this new protectionist movement. This region's politicians would form a powerful block in support of tariffs. In the

1850s, it would carry Lincoln into the White House and in the 1890s carry William McKinley to the presidency. It would be a Republican stronghold into the 1920s.

Clay would live to see the world hail his "American System" at the Great Exhibition of 1851. Queen Victoria and Prince Albert had designed the World Exhibition to be a coronation of Britain industrial supremacy, but it was the upstart Americans that captured the crowds in the Crystal Palace. In particular, the heavily protected textile, ordnance, and woodworking mills showed advanced machinery that challenged Britain's claim to be the world's premier manufacturer. Prince Albert formed a Committee on the Manufacturing of the United States. How had the world's best example of Adam Smith's free trade principles been trumped by the world's greatest protectionists? According to Adam Smith, protectionism would suppress innovation and invention. What had happened was the stabilized manufacturing market attracted capital investment and that stimulated innovation. The committee found that American manufacturing aggressively pursued experimentation and invention. Even worse the free trade environment in England had reduced investment in the manufacturing of machinery. Furthermore, since the 1820s, there had been a steady emigration of mechanics to the Untied States. Clay, not Smith, had correctly predicted that the application of free trade principles would move Britain from a great manufacturing nation to merchants. The same trend can be seen today in the United States.

Clay found support and hope in the textile factories of Lowell, Massachusetts. One of the deep fears of the Jeffersonians was that American cities would become the "Satanic mills" of England. The great textile mills of Lowell were much different places. They were positive and healthy working environments in the 1830s. Ventilation and light were abundant in the factories of Lowell. The communities offered education for the workingmen and women as well. The workers earned enough with the aid of company supported housing to save money. Clay would often send critics to Lowell to see the "American System." Things were not as utopian as advertised, but it was an dramatic improvement over British cities such as

Manchester. Lowell would even win over Daniel Webster to protectionism, and when President Andrew Jackson visited, he faced hundreds of banners declaring "Protection for American Industries." This is the model that H. J. Heinz, George Westinghouse, and George Pullman would hope to emulate decades later.

Clay found little presidential support in his lifetime, but his visits to manufacturing areas such as Pittsburgh, New England, the Mahoning Valley, and Cincinnati produced the largest crowds ever seen in American politics. Mills, factories, furnaces, mines, and streets (and many children such as Henry Clay Frick of Pennsylvania) were named after him. After his death in 1852, a young Abraham Lincoln took up the cause of the American System, and it remained the cornerstone of Republican administrations throughout the last half of the nineteenth century thanks to William McKinley. Lincoln used tariffs to raise money for the war, which was a basic use of tariffs as proposed by the Federalists. Lincoln's economic advisor, Henry C. Carey was a huge supporter of Clay's American System. Carey saw the free trade of the British as a great error, stating in his 1851 book, *The Harmony of Interests*:

> The whole basis of their [England] system is conversion and exchange, and not production, yet neither makes any addition to the amount of things to be exchanged. It is the great boast of their system that the exchangers are so numerous and the producers so few, and the proportion which the former bear to the latter the more rapid is supposed to be the advance towards perfect prosperity. Converters and exchangers, however, must live, and they must live out the labor of others: and if three, five, or ten persons are to live on the product of one, it must follow that all will obtain but a small allowance of necessaries or comforts of life, as is seen to be the case.

These words certainly ring true to the principles of protectionism. Carey became a key political ally of Clay, forming the Pennsylvania Society for the Encouragement of Manufacture, as well as the American Industry League. Carey was a prolific writer in support of tariffs throughout his career.

Carey requires some note because he was the most influential economist of the 1850s, 1860s, and 1870s. He was for easy money and strong tariff support; ideas supported by both Clay and McKinley. He understood the nature of the money supply as a stimulus, and supported the printing of greenback dollars. Today, of course, he would be considered an inflationist, but he argued while inflation hurt the bankers, it helped the manufacturers. He correctly identified the "enemy" as banking monopolists who favored importing and trade. Carey argued that these bankers were actually hostile to American industrial enterprise. Carey also predicted the bankers' takeover of the railroad industry to control trade. His writings foreshadowed the rise and dominance of J. P. Morgan in the McKinley era. Carey was the major influence on Lincoln's tariff policy that would become the policy of the Republican Party for many decades. Carey's disciples in the Congress such as "Pig Iron" Kelley and Thaddeus Stevens carried the protectionist banner in the period between Clay and McKinley.

CHAPTER SIX

The Rise of the Pig Iron Politicians

The pig iron industry dominated Pennsylvania, western Virginia, and Maryland in the late 1700s and early 1800s, so it is not surprising that it gave birth to a new group of politicians. These pig iron politicians stood out in a nation of farmers, representing industry and urban centers. In the constitutional ratification debate of the Pennsylvania state legislature, two embryonic parties started to evolve. These were the Federalists, who favored a strong central government, a national banking system, and excise taxes. The other party was an anti-Federalist party, known as "Republicans." The radical "Republicans" (Jeffersonian Democrats or Democratic Republicans) were concentrated in western Pennsylvania with the exception of Pittsburgh's Allegheny County, which the early Pig Iron Aristocracy held for the Federalists. These western counties were pig iron producers and heavily Scots-Irish, but excise taxes trumped all other issues in the rural areas. It would take decades for the Pig Iron Aristocrats to gain political control because of the political split of the Scots-Irish in the frontier iron regions over federal taxes.

Pennsylvania from 1810 to 1880 was the Pig Iron State, including eastern and western sides of the states. The western pig iron manufacturers increased after 1830, but the eastern part of the state maintained pig iron production and expanded its rolling mills. The Lukens Company outside Philadelphia pioneered the rolling of broiler plate in 1810, and would become critical to the evolving use of steam power. The eastern pig iron manufacturers remained an important industry factor into the late 1800s. The massive anthracite coalfields of eastern Pennsylvania were important to pig iron production as well. Pennsylvania became known for its tariff-supporting senators and congressman. The eastern politicians were less affected by the popularity of Andrew Jackson and stood firmly with the old Federalist beliefs.

French-Swiss Albert Gallatin (1761-1849) was one of the first of the pig iron politicians. While not an ironmaker, he represented the iron counties of western Pennsylvania in Congress. Gallatin had emigrated from Switzerland to Virginia in 1780. As an interpreter, he came to know many prominent politicians such as George Washington. In fact, it was Washington who suggested he buy land in the "Monongahela Country." Gallatin came to Fayette County, Pennsylvania, in the early 1790s, and built a large mansion on the banks of the Monongahela River. The Monongahela country at that time was probably 60% Scots-Irish and 30% German. The industries were the iron plantations, which also produced large quantities of "Monongahela Rye" whiskey for export. The Fayette iron district furnaces had strong ties and supplied James O'Hara and Isaac Craig of Pittsburgh as well as being the main contractors of pig iron for the United States government. Isaac Meason was the area's main industrialist, but there were other prominent men in the area such as General St. Clair. Most of the area's finances were tied to Pittsburgh banks, and Pittsburgh industries were the main customers. Pittsburgh's Bank of Pennsylvania functioned as a branch of the United States Bank, which was dedicated to economic expansion, and had James O'Hara and John Wilkins as directors. The bank's chief clerk was a young John Thaw, who through investments would become one of the area's richest pig iron aristocrats and create a family of pig iron aristocrats.

Albert Gallatin appears to have arrived in this country with some wealth, which he increased through investments in land and business. Gallatin had served on the Pennsylvania constitution ratification committee, in which he voiced opposition to taxes and strong federal government, an anti-Federalist position. Gallatin was elected to Congress in 1790 to represent Allegheny, Washington, and Fayette counties. This congressional district included Pittsburgh and was the heart of pig iron production and use. He was elected to the United States Senate in 1793, but Federalists held him out of that body based on the fact he had not been a citizen for nine years. In 1797, he opened one of the first glassworks west of the Alleghenies. Originally, the glass was made using wood as fuel, but it soon switched to the abundant outcroppings of coal at near Connellsville. Gallatin was a Jeffersonian Republican as were most of his Scots-Irish neighbors after the Whiskey Rebellion. He had been a protest leader in the Whiskey Rebellion, but helped sign a compromise in the end. In Congress, he continued his battle on the whiskey tax, becoming a hated enemy of the Federalists. Still, Gallatin was the first to propose a national road in congress in 1797. While a Jeffersonian Republican, he was of the unique pig iron branch of the party that supported Federalist national improvements and protectionism. Ultimately, decades later, this branch of the Jeffersonian Republicans and a similar branch in the Federalist Party would lead to the Whig Party and then the Republican Party.

Another key political figure of the Pittsburgh area was Hugh Henry Brackenridge (1745-1816). He was a Scots-Irish Federalist lawyer educated at Princeton. Brackenridge came to Pittsburgh in 1778, and he immediately became one of the city's greatest promoters. In 1887, he predicted that Pittsburgh would become "the greatest manufacturing center in the continent or even the world." This prediction was made when Pittsburgh was a small cluster of blacksmith shops with less than 1,500 people. Brackenridge was one the Scots-Irish torn by the Whiskey Rebellion; he had supported his cousins in the hills, losing social status in Pittsburgh. He was part of area's political split of the Federalist Party. Brackenridge sided with the liberal Federalists of Westmoreland and Washington County forming the Democratic-Republican Party in support of Thomas Jefferson. Brackenridge became the number one enemy of the pig iron Federalists. He did gain popularity among the hill Scots-Irish people and Albert Gallatin.

The Pig-Iron Aristocracy

From 1801 to 1813, Gallatin served as Secretary of the Treasury to Presidents Thomas Jefferson and James Madison. Jefferson appointed him because of his political ties in western Pennsylvania, which was strongly anti-Federalist. As his home voter base, Gallatin had a little Federalist in him. After convincing Jefferson to repeal the whiskey tax, he argued for a national road and a canal system. Western Pennsylvanians wanted no taxes or government interference in business, but did want roads and canals for commerce. Gallatin was not initially a big supporter of Jefferson's embargo against England, but its simulating effect on iron, glass, and cloth production changed his mind. Gallatin's political evolution reflected that of the Pig Iron Aristocracy. The poor Scots-Irish farmers as well as the wealthy Scots-Irish of industry hated the whiskey tax. Iron plantation owners always produced whiskey as an ancillary product, and urban industrialists such as O'Hara in Pittsburgh were in the distillery business. Monongahela Rye was actually bigger business than iron or glass in the 1790s. The wounds of the Whiskey Rebellion would dog Pennsylvanian politics until the 1830s. Federalist planks such as roads, canals, and banking had support with these frontier Jeffersonians. His support of the National Road was critical to its building in the 1820s. There was a split of the poor hill Scots-Irish and their capitalist cousins in the 1800s. The hill Scots-Irish remained stubbornly in the Jeffersonian and later Democratic Jacksonian camps.

The Whiskey Rebellion had been a defining event in the politics of the western frontier centered around Pittsburgh. It allowed for the rise of the Democratic-Republican Party of Jefferson to take roots in the area, since the excise tax was a Federalist measure. The next defining event was the War of 1812 and the economic warfare of Britain that continued into the 1820s. Having lost the war, Britain made a decision on a national level to dump iron, wool, and glass products in the United States. The great growth of the pig iron industry prior to and during the War of 1812 was crushed by this economic warfare of Britain. By 1816 most of the rolling, slitting, and nail mills of the tri-state pig iron industry were in or near bankruptcy. In the economic depression of 1816-1821, Pittsburgh's employment was cut in half. Congress had also decided to bypass Pittsburgh by building the National Road through Wheeling, which probably was due in part with Pittsburgh's role in the Whiskey Rebellion. The rise of the two-party systems had effectively checked the economic growth with a split in national unity. The economic depression of the period gave rise to "independent" tickets based on tariff and economic issues versus Federalist or Democratic-Republican leanings. This approach gave rise to the Senate career of James Ross of Pittsburgh. James Ross had the strong support of the Pig Iron Aristocracy as a Federalist. In addition, the City of Pittsburgh elected a "Select Committee" in 1816 to support growth regardless of party planks. Ross having the support of the "Pig Iron King," James O'Hara, headed this key non-partisan committee. Scots-Irish Ross brought in and allied with German industrialist and glassmaker Benjamin Bakewell. Glass was even more sensitive to imports than iron.

The early pig iron politicians reflected this ambivalence on a national level. Henry Baldwin, elected in 1820 to Congress, was a Federalist who switched to the Jeffersonian Republican (also known as the Democratic-Republican) party, but he was always a fierce protectionist. In the 1830s, the Democratic-Republicans split over protectionism. In fact, it was in this election that the Pig Iron Aristocracy of Pittsburgh put together a coalition that crossed party lines. This would also bring in Republican William Wilkins of the state assembly to this coalition. Eastern Pennsylvania politicians had formed such a coalition earlier to support its pig iron production. The coalition stood strongly for tariffs, and the recession of the late 1810s made tariffs popular to rejuvenate the pig iron and pig iron products industries. The dumping of pig iron by England became a rallying point. The industrialists were basically Federalists, but supported the coalition. The coalition was misled by Andrew Jackson supporters into believing Jackson was pro-tariff in the presidential election of 1828. These types of protectionist coalitions were formed in Cincinnati, Lexington, Philadelphia, Pittsburgh, and the Mahoning Valley. In the United States Congress the pig iron state Representatives formed one of the first caucuses. Generally, these pig iron districts included glass, coal, and other types of manufacturing. Tariffs were just as popular in the clothing industry of New England and eastern Pennsylvania. It could take Henry Clay in the 1830s to unite them nationally into a party. The rise of Andrew Jackson's popularity would cause splits until the formation of the Republican Party of Abe Lincoln.

The split in the Federalist Party was often seen more strongly in the Pittsburgh capitalists. Ebenezer Denny was a Pittsburgh capitalist with strong ties to men such as James O'Hara. He served with O'Hara on the board of directors of the Bank of Pennsylvania, and was his partner in a number of real estate deals. Denny was also a founder of the First Presbyterian Church and a member of the Masonic Lodge 45. In 1816, Denny became the first mayor of Pittsburgh. He like most of Pittsburgh capitalists had came through the military commissary corps and served at Fort Pitt. He was a veteran of the Revolutionary War, serving under General and Fayette County ironmaster Arthur St. Clair. He stood on the side of law in the Whiskey Rebellion even though his heart was with the farmers. Denny was typical of the pig iron politicians who lacked a political party that fully represented them in the 1820s. Their views would later be fused into Henry Clay's Whig party. James Ross was Pittsburgh's leading Federalist. Ross was a lawyer and United States Senator who moved to Pittsburgh in 1803. He was a partner and friend of James O'Hara. Ross was typical of the politicians more focused on local issues than national party politics. The Scots-Irish Ross was a member of Masonic Lodge 45 and the Presbyterian Church. As a politician, he was clearly controlled by the Pig Iron Aristocracy. One of his law understudies, William Robinson, would branch out in businesses such as bridge building, railroads, and banking, as did Ross. Ross, Wilkins, and O'Hara formed many Scots-Irish networks and partnerships, and all were active Federalists. Robinson would be involved in the formation of the Exchange Bank of Pittsburgh, and he would become the president of the Ohio and Pennsylvania Railroad.

The Pig Iron Aristocrats of Western Pennsylvania did take control of the state legislature. In response to the bypass of the National Road, they put together an alliance of state money and private money to invest in turnpikes. The Pennsylvania Turnpike from Philadelphia was already a well-developed road to compete with the toll-free National Road. The pig iron users of Pittsburgh, however, needed a direct connection to the pig iron furnaces of the Juniata Valley. The Harrisburg to Pittsburgh road known as the Huntingdon Turnpike passed through this critical pig iron valley. The Huntingdon Turnpike was completed in 1825 and paid for in a year, thanks to the traffic of Juniata iron to Pittsburgh. Round trip cost of a wagon delivering pig iron from the Juniata Valley was around twenty dollars based on four thousand pounds of pig iron in a wagon. The Pig Iron Aristocrats would force the state funding of the Philadelphia to Pittsburgh canal in the 1830s. While it helped the growth of the pig iron industry and was truly an engineering marvel, it was never profitable.

Pittsburgh's newspaper of 1829, the *Gazette*, was edited by Neville B. Craig, who was the son of Isaac Craig and Amelia Neville. The paper reflected the unusual politics of the Pig Iron Aristocracy and the region. The *Gazette* was called "conservative Democratic, Anti-Masonic, and Whig" all in one. The Anti-Masonic movement in western Pennsylvania was more an anti-Jackson movement, since it was being used nationally against Jackson's membership in the Masons. Both Neville Craig and Harmar Denny, who led the anti-Masonic movement, were sons of founders of Masonic Lodge 45 in Pittsburgh. This type of confusion also is an excellent example of the politics of the Pig Iron Aristocracy and the struggle to define their protectionist plank inside a national party. Eventually, the Pig Iron Aristocrats would unite as Republicans in the Whig Party. Initially the *Gazette* was considered Jacksonian in the 1830s, but its strong stand on protectionism brought it into the Whig Party of the 1840s and Republican Party of the 1850s. The common denominator of western Pennsylvanian and Ohio politics after 1820 was protectionism and industrial growth. It was as noted, a struggle against the Scots-Irish whiskey rebels in the population who personally voted Democrat because of the popularity and Scots-Irish heritage of Andrew Jackson. Neville Craig, editor of the *Pittsburgh Gazette*, became an active member of the Anti-Masonic Party. Craig hatred of the Masons put him at odds with Henry Clay as a presidential candidate. In 1842, the paper opposed Clay's presidential run because Clay was a Mason and slaveholder. In 1844, a group of Pittsburgh industrialists including Edward W. Stephens, William Eichbaum, and Thomas Bakewell helped implement a change in the *Gazette*'s editorial approach. From 1844, the paper became an advocate of Henry Clay and his "American System" of protection. This demonstrated the extensive political power of the Pig Iron Aristocracy in western Pennsylvania. Almost every President and presidential candidate had to call on the Pig Iron Aristocracy with a visit to Pittsburgh from 1830 on. Every president from Lincoln to Teddy Roosevelt visited Pittsburgh.

Pittsburgh had many visits from presidents and royalty between 1817 and 1900. Not counting George Washington's early visits, President Monroe made the first political presidential visit in 1817 to thank the young Pig Iron Aristocrats for their help in the War of 1812. At the time, Pittsburgh lacked a quality hotel, and President Monroe stayed at William Wilkins's home. In 1840, the Monongahela House was built as a first class hotel on a par with the best in the nation. Its guests would include eight Presidents: John Q. Adams, Andrew Jackson, Zachary Taylor, William H. Harrison, Abraham Lincoln, Ulysses Grant, William McKinley, and Theodore Roosevelt. It also included presidential hopefuls: Henry Clay, James Blaine, William Sherman and Philip Sheridan. Out of country guests such as King Edward VII stayed at the Monongahela House. The bar served only hard Scots-Irish whiskey, allowing those who drank beer to find a nearby German beer garden. Fresh oysters and sea fish were always on the menu. The Pig Iron Aristocrats tended to be well educated and loved the theater and the arts, bringing such artists to the Monongahela House as Mark Twain, Ralph Emerson, Horace Greeley, Henry Ward Breecher, and Thomas Benton.

The mix of Jacksonian Democracy and Whig protectionism was often seen in Pig Iron politicians in the 1830s. William Wilkins, a Jacksonian Democrat, as a senator and congressman voted with Clay's Whig Party and became known as the "Iron Knight." Yet, Wilkins was an independent thinker and an astute self-promoter, going against the Pig Iron Aristocrats and supporting Andrew Jackson's destruction of the national banking system, even though he had been president of the Bank of Pittsburgh. The National Bank had fueled the economic boom in the pig iron industries in the late 1820s. Anti-Jacksonian Congressman Harmar Denny would lead the local effort to save the Bank of the United States, since capital was the cornerstone of the pig iron industry. Denny led a group of Pig Iron Aristocrats to Washington to meet with President Jackson and protest the closure of the national banking system. The divided Jacksonian Democrats usually carried national elections, but Whigs such as state Governor Thaddeus Stevens carried local elections. From 1840 to 1853, Pittsburgh finally united the Pig Iron Aristocrats under the Whig Party. Wilkins was rewarded well by the Democrats with Jackson appointing him as Ambassador to Russia and serving as Secretary of War under President Tyler. The German families such as Eichbaum, Bakewell, Stephens, and Negley led the turn of the Pig Iron Aristocracy into the Whig and ultimately the Republican Party. Historically, the Germans of the iron districts had supported the Federalist Party, and never suffered from the political split of the Scots-Irish. The Pig Iron Aristocracy's political strength had always been in the alliance of German and Scots-Irish bankers and manufacturers. They had always favored the Federalist planks of a central bank and government support of industry. The Whigs actually, like Henry Clay, evolved out of the protectionist wing of the Democratic-Republicans. The Whigs incorporated high tariffs in those Federalist political planks at the beckoning of the Pig Iron Aristocracy.

    The mixture of pig iron and politics made Pittsburgh a natural banking center. Pittsburgh banking would be the root system of the Pig Iron Aristocracy, spreading out to western Pennsylvania, Wheeling, Youngstown, Cleveland, and southern Ohio. By 1870, Pittsburgh had over seventy banks, including sixteen national banks. The Pig Iron Aristocracy's tie with banking had started with James

O'Hara, continued in 1850s with James Laughlin, and grew into the 1870s with Thomas Mellon. The Pig Iron Aristocrats needed huge amounts of capital to build and expand their industry; no other industry had such huge capital requirements. The need of ancillary infrastructure such as canals, railroads, turnpikes, docks, and bridges also required large amounts of capital. In the 1830s, the Bank of Pittsburgh branch of the National Bank was dominant. Andrew Jackson's passion for busting the Bank of the United States helped to strengthen the Whig Party. Jackson started the move towards a national gold standard.

The gold issue was dominant after the Civil War resolved the slavery issue. The issue of gold versus silver, while as hot politically, lacked the clarity of slavery. The history, however, enjoys the same twists and turns as the early slavery issue. In the Coinage Act of 1873, the United States had a bimetallic policy. The policy allowed unlimited coinage of gold and silver with the restriction that the ratio be sixteen units of silver for every one unit of gold. Silver became undervalued in the ensuing years. In the marketplace, it took thirty-two units (grains) of silver to equal one unit (grain) of gold. This difference meant everyone wanted gold coins and certificates, it was more valuable. This hurt the silver mines in the west and to a degree reduced the money supply. The farmers with higher mortgage rates felt a reduced money supply. The Pig Iron Aristocrats wanted to increase the money supply to assure capital for investment.

Two other important politicians, James Blaine and Philander Knox, would be raised in the pig iron district of western Pennsylvania in later half of the 1800s. James Blaine came from a pig iron family with roots back to the Revolutionary War. While he settled later in New England, he was a staunch supporter of the iron industry and protective tariffs. He was dominant in the Republican Party in the period between Lincoln and McKinley. He was popular with the old "Iron Whigs" of the Mahoning Valley, and like William McKinley he picked up the mantle of Henry Clay and 'Pig Iron" Kelly. Blaine ran for president several times, missing by only a small margin. He served in both houses of Congress as Maine's representative and was Secretary of State under two presidents. He helped form the scientific protectionism of the Republican Party, and was well financed by the Pig Iron Aristocrats. In particular, he defined the principle of "reciprocity" between American trading partners, which years later William McKinley would incorporate in his tariff bills.

Philander Knox would be an important senator from Pennsylvania in the 1890s. He was a close friend of the Pig Iron Aristocrats, who often played poker with Henry Clay Frick. He was a Republican floor leader on tariff issues. He was brought in as an Attorney General for President McKinley and Secretary of State for President Theodore Roosevelt. He was often criticized for his support of Trusts in the late 1890s and early 1900s. He was a poker friend and business associate of Andrew Carnegie. He also was very close to the Ohio Pig Iron Aristocrats such as Mark Hanna. He was a fierce protectionist of the pig iron and steel industries.

Protectionism would ultimately be the issue of the Republican Party from 1850 to 1920, and the Pig Iron Aristocrats became the party core. Pennsylvania's "Pig Iron" Kelly and Ohio's William McKinley were the great politicians of the Pig Iron Aristocracy. They had both served as chairmen of the powerful Ways and Means Committee. William McKinley found national support like his hero Henry Clay in all pig iron producing districts. McKinley was the national apostle of protectionism, and had fought for tariffs throughout his congressional and presidential career. McKinley, like most Americans, was not clear on the gold issue. He wanted money supply growth and didn't care whether that was achieved by gold, silver, paper, or some combination. As "Jupiter" in the Pig Iron Pantheon, McKinley accurately reflected the view of the Pig Iron Aristocracy. In McKinley's presidential elections, he carried the pig iron districts of Youngstown, Pittsburgh, Cleveland, and Wheeling by an 18 to 1 margin! "Pig Iron" Kelly certainly influenced McKinley. For almost twenty years, McKinley was the protégé of Pennsylvania's Pig Iron Kelly on the House Ways and Means Committee. Kelly influenced McKinley in seeing that democracy and capitalism are interrelated in America; they combine to form the

"American System." Capitalism without opportunity, upward mobility, and civil rights is no better than tyranny. Kelly had also fought for the shorter workweek and better working conditions. McKinley incorporated all of this into his vision. It was Clay's system adapted for the Industrial Revolution.

The next political twist in the gold story came in the Coinage Act of 1873. The Panic of 1873 was successfully blamed by the bankers on silver in our coinage. Silver caused inflation, which is always the enemy of the rich. The Coinage Act of 1873 put the country on the gold standard. The "free silver" supporters knew it as the "crime of '73." The Panic of 1873 devastated the pig iron districts of Pittsburgh, Wheeling, Cleveland, and Youngstown. If you feel a little confused by whole silver/gold debate of the Gilded Age, don't. Confusion abounded in the Gilded Age between the average American and middle class voter. Voters could easily be manipulated by politicians, who themselves didn't understand the issue, but did the vote count. The issue split both parties. It came up in every election with continuing intensity until 1900. It wasn't until 1894 that free silver advocates realized they needed to educate the population. William Harvey published a little book called *Coin's Financial School*. It was to be as *Uncle Tom's Cabin* had been for the slavery issue. Harvey's book quickly became a best seller, reflecting the degree of confusion and the desire to understand. It also split the eastern bankers who wanted a gold standard and the Pig Iron Aristocrats who wanted more money supply for investment. Proirity one would be the Tariff Acts to protect iron. The McKinley Tariff Act of 1890 would be the high point of pig iron politics, but the gold issue threatened to split the Republicans.

In 1890, prior to Harvey's book, silver had strong support by the farmers, western states, small businesses, many manufacturers, and the Irish. Some congressmen saw an opportunity to use silver as wedge. The chant by Senate Democrats and western Republican Senators was "No silver, No tariffs." Congressman McKinley was now personally torn. The Senate was holding up his tariff bill as they debated free silver. McKinley was a bimetallist, even though there was a great deal of support for free silver in his district. He

The Rise of the Pig Iron Politicians

had opposed free silver often on the House floor, but now the tariff bill was clearly blocked. President Harrison threatened to veto any free silver bill, and the silver Republicans and Democrats didn't have the votes to override a veto. The result was the passage of the compromise Sherman Silver Purchase Act, which took the country off the gold standard. The Sherman Silver Purchase Act returned the country to the 16 to 1 bimetal standard. McKinley supported the compromise, so the Senate could move on to the tariff bill. Mark Hanna was outraged as were most of the Republican big money supporters. With the silver bill and anti-trust bills passed in July of 1890, the Senate moved ahead on the tariff bill. The McKinley Tariff Bill went to conference review in September, the conference committee faced rate reviews of 4,000 items and 450 bill amendments. The members, worn by the long House and Senate debate, passed it in seventeen days, and then-President Harrison signed it in October. The McKinley Tariff would usher in a great industrial boom in pig iron and steel production.

CHAPTER SEVEN

The "Iron Whigs"

The Ohio Wing of the Pig Iron Aristocracy

While Pittsburgh remained the iron-manufacturing center over Ohio, Ohio was the political heart and soul of the Pig Iron Aristocracy. The Ohio Pig Iron Aristocrats were of the same Scots-Irish Masonic stock of those of western Pennsylvania. Many, like the McKinley family, had emigrated from western Pennsylvania. It was in eastern Ohio that the Pig Iron Aristocrats would ally themselves with the Whig Party becoming the "Iron Whigs." Later this area would be the cradle of the Republican Party and protectionist center of American industries. The charcoal iron furnace industry of the early 1800s depended on iron ore, limestone, and most importantly, hardwood charcoal for fuel. Water power was also needed to power the bellows to raise the heat. All necessary elements came together in the Mahoning Valley and Hocking Hills area of Ohio. The first furnace in the Mahoning Valley was on Yellow Creek in 1802. The operation of a furnace was hard work. President William McKinley's father managed charcoal furnaces in the Mahoning Valley. The Scots-Irish were known to mix heavy drinking with ironmaking, but the McKinleys' conversion to Methodism kept them sober managers, much like the ironmaster in Dickens's *Bleak House*. The capital and knowledge came to Ohio from two sources, the English of Connecticut and the German and Scots-Irish of western Pennsylvania.

The mixing of politics and iron production had roots not only in Pittsburgh, Ohio's Western Reserve and Mahoning Valley. The rise of the Ohio Pig Iron Aristocracy can be traced through the family of President William McKinley and the Scots-Irish movement west. McKinley came from a long line of ironmasters, blacksmiths, and manufacturers with deep ties to the old Pig Iron Aristocracy. He was a true patriot, with two great-grandfathers serving in the Revolutionary War and a grandfather in the War of 1812. His father, William, was one of the industrious Scots-Irish who were also known as the "Ulster Scots." The McKinleys were Scots Highlanders who had been enticed to Northern Ireland in the 1600s to colonize Ireland for the British. Finding little religious and economic freedom in Ireland, they immigrated to America. The Scots-Irish were Presbyterians, who were often persecuted by the British because of their strict Calvinism and commitment to the Scottish Kirk. The Scots-Irish would also become the early industrialists of America. James McKinley, the future president's great-great-grandfather, came to a Scots-Irish settlement in the South as a blacksmith and ironmonger. On his mother's side his ancestors were Dutch blacksmiths, who escaped Europe with William Penn. The ancestors on his mother's side had an iron furnace around Doylestown, Pennsylvania, which produced cannonballs for the Revolutionary War. His great-grandfather, David McKinley, moved to western Pennsylvania with a group of Scots-Irish to start an ironmaking business. Western Pennsylvania was the frontier at the time, allowing the McKinleys to produce iron without the interference of the British Iron Act of 1750, which prohibited the colonies from producing iron products. George Washington's brother, Lawrence, also had a frontier iron furnace near western Pennsylvania. This area was rich in low-grade iron ore, and hardwood fuel.

David McKinley joined the many Scots-Irish iron-making settlements (known as iron plantations by the Scots-Irish) in Westmoreland County, Pennsylvania, east of Pittsburgh, after service in the Revolutionary War. Like most of Westmoreland County, David McKinley was a Jefferson Democrat. McKinley had been part of the Scots-Irish who participated in the Whiskey Rebellion of the 1790s. This area experienced a boom in 1812, as armaments were needed for the War of 1812. The British had suppressed ironmaking for centuries in America. The Scots-Irish responded as they always did to a lucrative opportunity, opening new furnaces in Pennsylvania and eastern Ohio. Immediately after the American victory in that war, the British began dumping cheap iron and textiles on the American market. While the British lost the military war, they imparted a crushing economic blow to American industry from 1813 to 1824. The flood of British goods resulted in the first recession in America, known as the Panic of 1819. Particularly hard hit was the western Pennsylvania iron industry. Baltimore editor Hezekiah Niles (the namesake of Niles, Ohio) teamed up with Pennsylvanian, Henry Carey to pursue Henry Clay, Speaker of the House, to bring action in Congress. The result was the Tariff of 1824, which was America's first protective tariff.

In the late 1790s, a number of Pennsylvania Scots-Irish ironmasters started to look to the Youngstown area. Both lines of McKinley's ancestors were drawn to the recently discovered iron ore deposits of the Mahoning Valley near Youngstown. This ore was known as "kidney ore," and although it was low grade, it would establish the historical roots of the Youngstown steel industry. The brown colored kidney shaped stones can still be seen in the valleys around Youngstown, particularly around Yellow Creek. The McKinley family came to New Lisbon, Ohio, to open an iron business. Furnace operations in the Mahoning Valley had started before the War of 1812. In 1803, Scots-Irish Daniel Eaton started the "Hopewell" furnace on Yellow Creek near the Mahoning River. This kidney or bog ore, while low grade according to today's standards, was superior to the ores of western Pennsylvania and thus pulled iron producers to the area. In 1806, another furnace was built on Yellow Creek by Robert Struthers at Struthers (Poland Township), Ohio, as pig iron demand increased with the deterioration of relations with England. Struthers had also emigrated from western Pennsylvania in 1797. Originally Struthers operated a gristmill, but the demand for pig iron due to the strained relations with England created a solid business opportunity. James Heaton began operating a furnace and forge on Yellow Creek near Poland, Ohio, prior to the War of 1812. Heaton was one of the first Americans to produce rolled iron bar, a product that would reign in the area for a hundred years. Gideon Hughes had also built a charcoal furnace at New Lisbon and partnered with the president's grandfather, James McKinley.

McKinley's father, William Sr., was born in New Lisbon in 1807. The McKinley family produced cannonball in the War of 1812 and cast iron products in peacetime in western Pennsylvania. William Sr. moved to nearby Niles in the 1830s, and leased the old Heaton Furnace in the 1840s. The President's parents, Nancy Allison and William McKinley Sr., married in Niles, Ohio. William Sr. was a manager and part owner of an iron furnace and rolling mill known as Rebecca Furnace. He also managed and helped build furnaces across the state line in Mercer County and Lawrence County, Pennsylvania. The family and the region depended on the tariffs that Henry Clay provided in order to keep their furnaces profitable.

The better technology, economies of scale, and government support had given the British iron industry major advantages. The furnaces of America still used wood for fuel while the British industry had converted to coal. Coal produced iron more economically and with higher quality. The struggling infant iron industry needed stability to experiment with coal as a fuel to make the necessary furnace adjustments. The Niles-Youngstown area had coal seams, and that would further advance experimentation. The Clay tariffs were just starting to generate the capital needed for such research and development in 1840s. David Tod started some of the successful coal mining operation in the Mahoning Valley. William McKinley Sr. (father of the President) teamed up with the famous ironmaster Joseph Butler Sr. to build Temperance Furnace across the state border in Mercer County, Pennsylvania in 1838. The furnace was later renamed "Harry of the West" in honor of Henry "Harry" Clay. The area would become the stronghold of Henry Clay, the Iron Whigs, and the Pig Iron Aristocracy. Joe Butler Jr.

would become a leader in the iron and steel industry, a backer of friend William McKinley, and a leader politically of the Pig Iron Aristocracy. Butler would become president of the American Pig Iron Association. Another nearby furnace (Fremont Furnace) employed the young future President McKinley. Joseph Butler Jr. would be a key agent in every presidential election from Lincoln to Taft. Butler often represented the pig iron industry before Congress. He would be remembered for his friendliness, but he was also the Pig Iron Aristocracy's most powerful ambassador.

Butler and his father often teamed up with David Tod and James Ward, Sr. in furnace building. Niles, Ohio, also had a secondary boom in the 1830s, as a very rich layer of black iron ore was discovered at Mineral Ridge. The discovery is attributed to James Ward, a highly skilled English ironmaster. Ward should be considered America's first iron metallurgist. Using a small furnace in New Lisbon in the late 1830s, he experimented with various ores and the properties of the iron produced. James Ward, his brother William Ward, and Thomas Russell moved their company to Niles to build a furnace and three puddling furnaces. Niles soon became not only a smelting town, but it grew its rolling mills for iron and steel. The Mahoning Valley was probably far ahead of Pittsburgh as an iron producing center in the 1840s due to the ore deposits of the valley and its canal system linking it to Cleveland. Ward was not only a furnace wizard but an expert metal worker. In 1841, he built the first bar rolling mill, west of Pittsburgh. The mill produced bar iron for blacksmiths, horseshoes, sheet, and plate. James Ward also invented a type steel and iron mix known as "Dandy Tire," for use on wagon wheels. This "mixture" was probably medium carbon steel that exhibited superior wear and welding properties for wagon tires.

The politics of iron came to play in the 1830s as the iron industry had a re-birth. Both Whig and Democrat United States Congressmen from Ohio voted for the strongly protective Tariff of 1832. The area was becoming strongly anti-Jacksonian because Jackson opposed protective tariffs, canals, and banks, which were all needed for the iron industry. It was here that the pig iron aristocrats of Mahoning Valley took leadership from the Pittsburgh aristocrats in coming over to the Whig Party. Canals were the hot issue of the late 1820s and 1830s. The Mahoning Valley was rich in coal but lacked the ability to move it economically to Cleveland and Pittsburgh. The same was true of pig iron. The Pig Iron Aristocrats of Ohio and Pennsylvania got state approvals in 1828, but business conditions prevented the start of the Pennsylvania and Ohio Canal until 1838. The canal would connect to the Ohio River at Beaver, Pennsylvania, and to the Ohio Canal at Akron. The Ohio River connection would allow a direct water connection to Pittsburgh and the Akron Ohio Canal connection would allow a direct water connection to Cleveland. Pittsburgh and Cleveland would also be connected. These connections would secure Youngstown as a future iron-making center. From Cleveland the connection could be made via lake to the Erie Canal. This connection would be the future path of the immigrant Carnegie family from New York to Pittsburgh in 1848!

The opening of the canal would create another Youngstown industrialist and politician, David Tod. Tod was one of the first to profit from the canal. He started a coal mining operation on his "Brier Hill" farm. The coal was shipped to Cleveland to be used as fuel. Tod also started experiments in the use of this coal and coke processed from it at Youngstown's Eagle Furnace. Eagle Furnace had been built in 1846 as a charcoal furnace by pig ironmaster Jonathan Warner. The furnace experiments showed his Brier Hill coal could work exceptionally well without being coked. This was the only coal known to have such properties. In the 1850s, Eagle Furnace set a world record of twenty-eight tons in one day. Eagle Furnace would bring industry leadership to the Mahoning Valley. Warner took Eagle Furnace to the forefront with Tod's Brier Hill coal and the new discovered "black-band" iron ore mixed with Lake Superior ore. The high quality pig iron produced here became known as "Warner's Scotch Pig." Eagle Furnace was the first to use air heating stoves for the blast as well, further improving its productivity. The success of the Youngstown blast furnaces made David Tod a very wealthy man. In 1861, Tod built his own iron furnace (Grace I) and formed Brier Hill Iron and Coal Company. Tod had been a Democrat that favored protectionism, canal building, and many of the Whig planks. He lost several political campaigns because of his affiliation with the Democrats. Finally, running as a "Union Democrat" supporting the Civil War and protectionism, he was elected governor of Ohio.

The evolution of the modern blast furnaces has its roots in the Mahoning Valley. Many consider Eagle Furnace's use of lake ore and Brier Hill coal the end of the charcoal furnace era in the late 1840s. The Eagle Furnace added the classic heating stoves seen in modern furnaces. Alexander Crawford's Phoenix Furnace built in 1854 was the first to use a skip hoist to add coal and iron ore, freeing furnaces from having to be built in hillsides. The new technology brought a number of great early blast furnaces to the Mahoning Valley by 1872. In 1872 the Mahoning Valley had twenty-one furnaces with the total daily rated output of eight hundred tons, more than any other district. Most of these furnaces were merchant stand-alone furnaces that supplied a number iron users. The rolling mills rivaled those of the Pittsburgh district. The Pig Iron Aristocrats dominated the Mahoning Valley far beyond those of the Pittsburgh district.

Many were comparing the Mahoning Valley's manufacture of iron products to America's earliest manufacturing center in the Connecticut Valley. In 1853, the first shipments of rich Lake Superior ore arrived in the Mahoning Valley via Cleveland and the northeast Ohio canal system. Lake Superior ore would change the iron industry forever. The canal system had been the national priority of Henry Clay, and a key cause of the area's industrial development and the influx of lake ore. The McKinley family had a series of small furnaces in the Mahoning Valley from the early 1800s. McKinley's father and Niles would struggle through the economic depression known as the Panic of 1837. Money for investment dried up and demand for iron products dropped dramatically. This depression and period of bank failures would make Niles a center for Henry Clay's Whig party. The Ohio canal system helped the area out of the depression by allowing new product markets.

In the early 1840s, the Mahoning Valley ironmakers made a breakthrough in blast furnace technology by using the newly found coal deposits. These coal deposits were found near Niles on Mineral Ridge. The technology saved the Mahoning Valley industry from extinction due to the lumbering out of the hardwood forest. It also created a new mining industry in the area. Niles, Youngstown, and Columbiana County all prospered with the use of coal in blast furnaces. The area also started to pioneer rolling mills producing bar stock and plate. In addition, the mills started to produce nails, which were badly needed on the western frontier. The eastern markets grew with canals, since wagon train shipment of iron over the mountains was difficult and costly. Blast furnace iron made by coal or coke was cleaner and stronger, giving the Mahoning Valley an edge in the world market. It would be in Niles that the future President McKinley was born.

John Tyler was elected president in 1842 as a Whig Party candidate. It was the same year that Karl Marx and Friedrich Engels met and started the socialist movement, which would be a factor in President McKinley's assassination. Henry Clay's Whig Party was in control with Whig President John Tyler in the White House. The McKinley families were Whigs and abolitionists. The Whig Party stood for abolition, a federalist approach to government, a strong banking system, and growth of American industry. But the Whigs' rise to power occurred with their support of protective tariffs, a national bank to supply investment capital, and a strong network of roads and canals to support commerce. The Mahoning Valley of Ohio was a Whig (and future Republican) stronghold.

The Ward family would come to dominate the Mahoning Valley. In 1841, James Ward had established the first rolling mill west of Pittsburgh in the Niles, Ohio, area. James Ward and Company built an iron furnace near Youngstown at Niles and began using local coal. Ward developed the most advanced furnace operations in America. Ward also improved on the hot blast to improve furnace yield. Puddling furnaces, pig iron furnaces and rolling mills coupled with the Clay Tariff of 1842 created an industrial valley. Prior to the Ward operation, pig iron was being shipped to Pittsburgh to be processed. While canal connections were being developed by Pittsburgh capitalists, most of the pig iron was being moved the sixty miles or so by pack horses. Rolling mills opened in the Mahoning Valley and they started to compete with those of Pittsburgh. These Mahoning Valley mills had a direct local source of pig iron. The Mahoning Valley of the 1840s was also becoming a pig iron source for Pittsburgh foundries and rolling mills. The Pig Iron Aristocracy was becoming a brotherhood of manufacturers that crossed regional lines.

The Brown family was another branch of the Ohio Pig Iron Aristocrats, which could trace its lineage to the early iron plantation owners. Joseph Brown was the family patriarch, who grew up on an old Pennsylvania iron plantation. He ran an iron furnace in Franklin County, Pennsylvania, in the 1830s. He learned every phase of ironmaking, coming to Youngstown in the 1840s. In 1843, he bought a small steel plant in Youngstown on the canal, and grew it into an integrated iron operation. While Pittsburgh Pig Iron Aristocrats were financially based, the Ohio aristocrats were operators and technical men. Brown advanced blast iron technology in the 1850s, and was known nationally throughout the industry. He became an industry leader and advisor to politicians. He was often used by "Pig Iron" Kelly and William McKinley to testify to the Ways and Means Committee on tariffs. Until his death in 1886, he served as president of the National Tariff League of America. It was Ohio pig iron aristocrats such as Joseph Brown and Joe Butler that made things happen in the United States Congress.

Another region of Ohio that challenged the Mahoning Valley iron supremacy was the Hocking Hills or Hanging Rock iron district that took in southern Ohio and parts of Kentucky. The name Hanging Rock comes from a sandstone bluff on the Ohio River. The area of rich iron ore totalled about 1,300 square miles, covering the Ohio counties of Lawrence, Scioto, Gallia, Jackson, Vinton, and Hocking. Enterprising Scots-Irish ironmasters came from Pennsylvania, Virginia, and Maryland to start charcoal furnaces, the first being Union Furnace in Lawrence County. Many, like ironmaster John Campbell, eliminated Sunday labor based on their strict Presbyterianism. The area was still frontier with abundant hardwood available for charcoal furnaces, which increased to over 50 furnaces by 1860. The main product of the Hocking Hills or Hanging Rock region was pig iron to be shipped to Cincinnati and Pittsburgh via the Ohio River. The first furnaces in the area had started with the Scots-Irish pushing west to avoid government, since whiskey was part of the business. The first major furnace of note was Union Furnace in 1826. The area was blessed with iron ore, virgin forest, and a national transportation system via the Ohio River. Most of the furnaces were within fifty miles of the Ohio River, but teams of fifty oxen were needed to get to the river. The iron ore, while low grade by today's standards, was rich compared to Pennsylvania and Ohio veins. The deposits ran in four- to eighteen-inch-thick veins and were easily mined.

By 1845, Hocking Hills had exceeded the pig iron production of Mahoning Valley, but lagged in forges and rolling mills. Hocking Valley companies did start some direct casting of cannon, cannonballs, farm implements, and stoves. The British considered the low silica ores of Hocking Hills to be superior in armament casting. The region exported large amounts of pig iron to England during the Crimean War (1854-1859). During the Civil War, Hocking Hills iron was in high demand with the cannon foundries of Pittsburgh. The Union Navy specified that Hocking pig iron be used in cannon and armor. In addition, the Hocking Hills Hecla Furnace cast the famous cannon, the "Swamp Angel," used in the siege of Charlestown, and Jefferson Furnace cast plates for the *Monitor*. During the Civil War, the region operated more than fifty furnaces and used over 10,000 acres of forest a year each, or about three hundred acres per furnace per year. Raw ore was also being shipped to Wheeling and Pittsburgh furnaces in the 1850s and 1860s. Even during the Civil War, the Hanging Rock furnaces were more reminiscent of the iron plantations of the 1700s. They were generally family-owned farms and distilleries as well as furnaces. The furnaces were simple cold blast charcoal furnaces, but at forty to fifty feet tall, they were the largest in the United States. During the Civil War, their profitability and productivity were unequaled. A typical Hanging Rock furnace could make eight tons per day at a cost of thirteen dollars per ton, and commanding a sales price of sixty dollars per ton because of the high quality. Quality was achieved through the purity of the local iron ore. The area was originally Jeffersonian Republican and Scots-Irish, but was converted to the protectionist planks of the Whig Party. Eventually these counties would be part of the Ohio Republican stronghold of the later 1800s.

Mahoning Valley had one advantage over southern Ohio and Pittsburgh. It had a very special coal deposit, which allowed for the rise of the modern blast furnace. The coal deposits in the area possessed some rare blast furnace qualities. Only the coal deposits around Glasgow, Scotland, had similar properties. This "splint" coal was a semi-anthracite coal capable of firing a blast furnace directly. Charcoal furnace stacks were limited to about forty feet, but the strength of coal allowed stacks to go much higher, to seventy or eighty feet. The higher stacks allowed more furnace output. The additional heat coupled with stack height allowed these coal blast furnaces to double the output of a charcoal furnace. Most higher stack blast furnaces in Pittsburgh and West Virginia used coke from coal at higher cost. The immediate cost of coking and transporting coke was eliminated with the use of Mahoning Valley hard coal. In the late 1850s, YoungstownYoungstown blast furnaces were leading the world. Mahoning Valley pig iron was supplying the rolling mills of Pittsburgh and Wheeling as well as creating new rolling mills in eastern Ohio. Some of this coal was also being used in southern Ohio's Hocking Hills district. Still the number of charcoal furnaces in southern Ohio allowed it to maintain the larger pig iron tonnage overall. Just as important was the education of the "wizard of blast furnaces," Julian Kennedy, in the Mahoning Valley.

The Cleveland-Mahoning Valley district did overtake the Hocking Hills district in 1873. The Cleveland-Mahoning district benefited by the discovery and shipping of richer Lake Superior iron ore into Cleveland by the lakes and Mahoning Valley via the Ohio

Canal and expanding rail service. The Mahoning Valley had better technology with the use of local coal, while Hocking Hills remained a charcoal based industry. Cleveland, in particular, grew rapidly in pig iron production from lake ores. The pig iron boom doubled the population of Cleveland as foundries, furnaces, and rolling mills employed thousands. Pig iron and related products became the basis of the Republican Party as the old "Iron Whigs" joined because of Lincoln's strong protectionist tariffs. More importantly the Mahoning Valley Pig Iron Aristocrats offered the bridge between other coal and iron using districts. Pittsburgh investors were also being brought into the Mahoning Valley, further strengthening the keystone role of the area.

In 1860, in Portsmouth, Ohio, the ironmasters held a convention more political in nature than technical or financial. The ironmasters threw their support to the Republican candidate Abe Lincoln. The Ohio Democrats started to use the term "Pig-Metal Aristocracy" to label these Republican supporters. The Pig Iron Aristocrats of Ohio and Pennsylvania gave the Republicans and Lincoln national and local victories based not on the slavery issue but on the tariff issue. The Pig Iron Aristocracy more than any other political block would help end slavery, although it was more an indirect result of their tariff support. The strong support of protective tariffs rallied the Pig Iron Aristocracy, and Lincoln carried every iron-making district in Ohio. The Lincoln election represented a new alignment in the politics of ironmaking. The Republican Party would eventually completely adopt the original American System of Henry Clay. It is completely factual that the "Pig Iron Aristocracy" would be responsible for Republican supremacy from 1859 to 1929.

The man who, more than any other, made the Pig Iron Aristocracy was Ohio's William McKinley. His presidency represented the height of power for the Pig Iron Aristocrats, and his death started the decline. There were, of course, many eulogies in the Congress at the death of McKinley, but the one that McKinley would have liked best came from his friend Senator Chauncey Depew:

...with Washington and Hamilton, with Webster and Clay, he came, not alone, as they did, by the cold deductions of reason, but also by observation and experience, to the conclusion that the solution of our industrial problems and the salvation of our productive industries could only be had by the policy of a Protective Tariff. As Union and Liberty had been the inspiration of his courage and sacrifices as a soldier, so now America for Americans became an active principle of his efforts as a citizen.

The great iron industry of America depended a lot on the iron merchants of Ohio's Western Reserve. These iron merchants formed a major link in the iron industry geographically and politically. The area around Cleveland and east was known as the "Western Reserve" of Connecticut, from the state's claim on the area in the 1700s. The roots of the Western Reserve can be traced to the Pilgrims of New England. Like all immigrants, the Puritans moved west to find prosperity. One of the earliest of these movements was to the mineral-rich Connecticut River Valley in 1636. The Connecticut River starts in the mountains of New Hampshire and cuts through Vermont, Massachusetts, and Connecticut into Long Island Sound. In the early 1600s some Pilgrims and Dutch had a fur trading operation, but not farming. It was an extremely fertile valley, which drew immigrants from New York and Pennsylvania, but farming would be only a base. More importantly, the Connecticut River was a huge energy source. The Connecticut River's fall over its length is estimated to have had an energy potential greater than that of Niagara Falls. The river could support power for gristmills and iron furnaces, making farming more economically successful. The power of the Connecticut River caused a mixing of nationalities, and the emergence of American manufacturing. The Valley would give birth to manufacturers such as Eli Whitney, Samuel Colt, Francis Lowell, Cyrus Buckland, Thomas Blanchard, Alexander Holly, and George Westinghouse. The ethnic mixing retained the Protestant work ethic and economic drive of the Puritans. The Connecticut Valley would launch the Machine Age in America and bring iron technology to Ohio's Western Reserve.

The Puritans in Connecticut expanded their crafts and skills in metalworking. Twenty-six towns recorded silversmiths before 1776. Silver plate was one means of storing wealth in a pragmatic manner, so favored by the Puritans. The demand for silver product created world-class silversmiths in New England such as Paul Revere of Boston. The Puritans of Connecticut really started the metals trades of America. Even with limited efforts to produce pig iron, the Connecticut colonists developed a secondary metals industry unequaled in the other colonies. Connecticut developed eight rolling mills (two more than the total of all the other colonies), which manufactured hoops for barrels, nails, and sheet for tinplate. The demand for iron bar created a search for bog iron, which was found in 1734 in Salisbury Township near the New York border. Initially, these industrious colonists used crude Catalan furnaces and small forges. By 1740, charcoal furnaces were being built; some of these became "iron plantations," making an array of products. Connecticut ironsmiths advanced the technique of steel making for small tools, knives, and banquets. The area became known as the "Arsenal of the Revolution." In 1762, Ethan Allen, future Revolutionary War hero, built his Lakeville Furnace. He added some of the large forges in America to make cannons and anchors. Without the Salisbury iron works, it is doubtful that the colonists could have armed properly for the British. After the Revolutionary War, these skilled metalworkers and traders formed the Connecticut Land Company to expand into northern Ohio.

The Cleveland east area became the destination of New England Puritans moving west, and with the immigration came iron technology. One of the earliest to settle the Western Reserve was Samuel L. Mather, a major stockholder in the Connecticut Land Company. Mather, out of Cleveland, explored the iron fields of Michigan in the late 1840s. In 1850 with a group of Cleveland businessmen, Mather organized the Cleveland Iron Mining Company. Mather was a true visionary of the future of iron in America. In the late 1850s, he opened an iron mine at Ishpeming, Michigan. Cleveland Iron Mining started to move ore to the Cleveland docks and then by canal to the furnaces of the Mahoning

Valley, Pittsburgh, and Wheeling. The first use of Lake Superior ore in a Sharon, Pennsylvania, furnace was by David Agnew in 1853. Agnew was an ex-Wheeling furnace man (originally Agnew had been a partner of Pittsburgher Peter Shoenberger), and the success of Superior ore spread quickly to Wheeling and Pittsburgh. Mather had built a huge ore transportation network from Michigan as the lower grade ores of Ohio and Pennsylvania were declining. The gathering point for these ores would be the Western Reserve of Ohio.

Part of this success was the development of a new ore boat with the navigation quarters in the bow. Cleveland docks handled 200,000 tons of ore in 1865, but the new shipping advances pushed that over 500,000 in 1868. Meantime Mather opened charcoal furnaces in Michigan to produce pig iron for Cleveland. The hardwood forests, however, could not support the large furnaces for long, and there was no coal in the north.

The next advance in lake iron ore would be in logistics. James Pickands and Henry Pickands, working out of Cleveland, started to receive ore and then ship coal to Michigan, allowing two-way commerce. Their father had run a hardware store in northern Michigan to supply the mines and miners. James Pickands became the manager of Michigan's Bay Furnace in 1868. In the 1870s, Mather and Pickands came together to form the iron ore company of Pickands, Mather and Company. This company became a mining, ore processing, transportation, and blast furnace operation. Cleveland would start its development as a pig iron producer thanks to Pickands, Mather and Company.

The real birth of the Ohio Iron Whigs was not in Ohio, but in Michigan and Minnesota. The massive iron ore deposits of Minnesota and Michigan had been known since the early Jesuits explored the area in the 1600s. Copper finds had caused a rush of speculators into Minnesota in the late 1830s. They found large iron ore deposits, and the vast hardwood forests made pig iron production a natural. The first charcoal furnace started production in 1847 at Carp River about ten miles from Marquette. In the 1870s, charcoal furnaces were operating throughout northern Michigan and

Minnesota. The pig iron was then shipped to Chicago and Cleveland. From Cleveland, pig iron and sometimes iron ore moved to the Mahoning Valley and Pittsburgh. The rich forests, however, were quickly being lumbered out for charcoal and for America's lumber and construction industry. In the 1870s, Pickands, Mather and Company expanded and developed iron mines in Minnesota. The company built railroads in the north to further improve mining and ore transportation. The transportation system to bring lake ore south would meet the coal mining advances to revolutionize the pig iron industry.

The final legacy of the Ohio Iron Whigs and the Pig Iron Aristocrats would be the great pig iron boom of the 1890s. This pig iron boom resulted from the emergence of America's tinplate industry after the passage of the McKinley Tariff of 1890. McKinley's Pig Iron Aristocrats proved him right with a huge investment in tinplate mills after tariff protection was enacted. The first large tinplate mill would open in McKinley's hometown of Niles in 1892.

CHAPTER EIGHT

King Coal

Coal and iron have been almost inseparable since the 1800s. It was the pig iron industry that caused coal mining to boom. The coal sources of Pittsburgh made it the "Iron City" even though it lacked any iron deposits. Coal was the source of iron production at Wheeling, Youngstown, and even Chicago. Coal not only replaced the diminishing hardwood reserves in charcoal furnaces, but also was the fuel used to reheat and melt pig iron for foundries and rolling mills. Coal rode the most powerful steam engines of the time to power the iron rolling mills. Coal and its product coke changed the nature of the iron plantations in western Pennsylvania, turning them into iron manufacturing factories. Coal not only made Pittsburgh the "Iron City" but also the "Glass City." James O'Hara's early glasshouse used coal as a fuel. Coal kings such as Thomas Mellon and Henry Clay Frick would further add financial and political might to the Pig Iron Aristocracy. Without coal, the pig iron industry would have died out in Pennsylvania and Ohio by the start of the Civil War. The history of coal is fundamental to understanding the evolution of the Pig Iron Aristocracy.

The British coal industry was well established in the 1700s, while America's rich forests made wood the preferred fuel. Coal was probably used at British forts on the colonial frontier in the 1750s. The early beginnings of coal mining in America brought skilled coal miners from England, Wales, and Scotland. Coal mining in the eighteenth century was considered a skilled trade. The famous "Pittsburgh" seam along the Monongahela River was at most six feet thick. That was much smaller than British seams. Cornish and Welsh miners had developed a technique using a dog sleigh to pull coal out of small mine holes. In parts of Western Pennsylvania these old holes can still be found. This type of mining was common around Pittsburgh and eastern Pennsylvania in the early 1830s, and was done by skilled miners (colliers). These miners were paid based on the bushels of coal they dug. Pittsburgh's Scots-Irish found industrial uses for coal, such as in glassmaking, which created a coal boom. Scots-Irish banker Thomas Mellon would make his fortune investing in western Pennsylvania coal mining. The demand for coal created the need for more miners and mining wages increased proportionately. Mellon and the Corey families had large coal mining and shipping operations at Braddock in the 1840s.

Pay more than anything drew British immigrants to the American mines. The following is from a letter of a British immigrant miner in 1819:

> Coal cost 3 cents per bushel to be got out of the mines. This price, as nearly as I can calculate enables the American collier to earn upon average, double the number of cents for the same labor that the collier in England can earn; so that as the American collier can, upon an average buy his flour for one third the price that the English collier pays for his flour, he receives six times the quantity of flour for the same labor.[13]

Economic motivation had been the root of immigration since colonial times.

American coal, like other American industries, broke down the crafts model of Europe because American demand required mass production. As a European craft, coal miners spent years learning their trade. As a trade, the miner set his own pace and work conditions. The early immigrant British miners brought the same crafts system to American coal mining, but the craft would change

---

[13] Carmen DiCiccio, *Coal and Coke in Pennsylvania* (Harrisburg: Pennsylvania Historical Commission, 1996).

as it evolved into an American industry. Cornish miners dominated the Monongahela Valley's small mines prior to 1800. Similarly, in Ohio's Mahoning Valley, Cornish and Welsh miners were operating small mines. These mines were often called "gopher holes," or "country holes." The apprentice system was firmly in place prior to the Civil War, when the Scottish, Welsh, and English dominated the mines. Some thirty-seven thousand British coal miners immigrated to America in the 1850s. These early British immigrant miners worked as sub-contractors. They were paid by the bushel and could set their own hours. Later, as cheap Irish laborers entered after the Civil War, the crafts system broke down.

As in other industries, the breakdown of the crafts system in the coalmines was a combination of consumer-driven demand and labor shortages. In the 1700s, Pennsylvania blacksmiths mined their own coal for their hearths. By the 1790s, the Ohio and Monongahela River valleys were using coal to heat houses because of its regional abundance. By 1810, Pittsburgh had become known as the "Smoky City" from home heating alone! Coal sold at six cents a bushel and burned better than walnut. The Scots-Irish of the region started to operate small mines to make money. These part-time miners were known as "winter diggers," "pumpkin rollers," "wheats," and "greenies." The abundance of coal in western Pennsylvania and eastern Ohio led to experimenting in many applications for its use. Cooking failed because of the odor, but blacksmith shops expanded it for a number of industrial applications. Pittsburgh was the first to use it in glassmaking in the 1790s, which for centuries had been based on wood. Isaac Craig, the Pittsburgh Scots-Irish industrialist, pioneered the use of coal in glass kilns. Even early iron plantation owners often manufactured glass as well. Pig Iron Aristocrats such as Craig started in glass manufacture. By 1837, the Pittsburgh glass industry was the largest in America and consumed 1.5 million bushels of coal a year, while Pittsburgh's iron industry consumed about 100,000 bushels a year.

Improvements such as puddling and coal-fired furnaces advanced the production of wrought iron. The wrought iron industry had been growing with the technological breakthrough by Scots-

Irish ironmaster Isaac Meason, the iron baron of Pennsylvania's Fayette County. Meason invented the puddling process to convert pig iron into wrought iron, which was needed for nails. A puddling furnace used coal heat to re-melt pig iron. Coke and coal were both used in these puddling furnaces, which came from demand for nails. By 1820, Pittsburgh had numerous nail mills for the ever-growing demand of the West. The puddling furnaces required massive deliveries of coal and coke from Fayette and surrounding counties. It would be Pittsburgh's transportation ties and nearby coal seams that made it the "Iron City."

Coal was gathered at ports such as Braddock, Beaver, and Brownville, Pennsylvania and then was shipped down the Monongahela River to points as far south as New Orleans. The 1820s saw the rise of coal usage in salt production. In the hills of western Pennsylvania, salt was evaporated from brine using coal as a source of heat. By the early 1830s, salt works were consuming 72,000 bushels annually. Consumption rose rapidly as steam engines replaced horses in pumping brine from the mines. The demand required more British miners to immigrate to western Pennsylvania. More farmers were also drawn into the booming market of the 1830s. The mining, processing, and shipping of coal created America's first masters of capitalism and captains of industry. In particular, it made many Scots-Irish immigrants extremely wealthy capitalists such as Thomas Mellon. King Coal would soon, however, be replaced by the coal needs of the Pig Iron Aristocracy. Henry Clay Frick would be one of the capitalists to bridge coal and ironmaking. Frick started in coal and rose to be Andrew Carnegie's partner. Leaving Carnegie, he partnered with J. P. Morgan and became one of America's wealthiest industrialists.

The real demand boom for coal was yet to come in pig iron production. The charcoal iron furnace had dominated the American scene until the 1870s. The problem in America was not coal technology, but the abundance of cheap wood for pig iron furnace fuel. The breakthrough technology had actually been found in 1735 at the British ironworks of Abraham Darby. Darby had used coke in his blast furnaces, coke being a coal-based product. Darby showed a major reduction in cost by the use of coke to make iron, but ironmakers were slow to adopt coke. Still, every year the shortage of wood in Britain became more pressing. On average an iron blast furnace consumed an acre of woodland daily. As wood ran out locally, furnaces switched to coke. By 1788, ninety-four percent of Britain's iron furnaces were using coke, while there were none in America.

British immigrant ironmasters were aware of the success of British furnaces in using coal, but wood was plentiful and there was financial motivation. Scotland was also exploiting its anthracite coal deposits. The scientific philosophers in America were aware of potential shortages in the future. In 1824, the scientific Franklin Institute of Philadelphia offered a gold medal to the first furnace to use twenty tons of iron from coke or bituminous coal only. The Bank of Philadelphia offered a cash reward of five thousand dollars. A group of British immigrant ironmasters went even further in 1837, using anthracite coal, which was plentiful in eastern Pennsylvania. This created a boom for anthracite, and a decade later western Pennsylvania began using coke from bituminous coal. The coal prince who would become a key part of the Pig Iron Aristocracy was Thomas Mellon, who would later finance coal baron Henry Clay Frick.

Eastern Pennsylvania coal was the hard, energy-rich anthracite, which could be used directly in iron furnaces. By the 1830s, most eastern Pennsylvania furnaces were using anthracite as fuel. Western Pennsylvania had soft, volatile bituminous coal, which needed to be converted to coke. Coke making, like charcoal making, was the slow burning of the coal for five to eight days. The coking process is technically known as "destructive distillation" because volatile oils and gases are burned off. The first experimental use of coke in an American iron furnace is attributed to Alleghany Furnace in 1811 at Blair County, Pennsylvania. The shortage of wood had forced the use of coke as early as 1735 in the famous furnaces of Abraham Darby. The same thing happened at Clay Furnace, Mercer County, Pennsylvania, which was forced to switch to coke in 1845. A few years later, Clay Furnace switched to "splint" coal from

Youngstown, which was a natural type of coke. Coke not only increased the heat of an iron furnace, but also supplied strength, allowing for much higher furnace stacks. Higher stacks meant higher production. Coke making also burns out sulfur in the coal that can greatly reduce the properties of the pig iron produced. Coke also made an excellent fuel for the re-melting of pig iron in foundries and puddling furnaces. Isaac Meason started using coke in his Fayette County, Pennsylvania, rolling mill and foundry in 1818. Generally, however, charcoal furnaces dominated until local wood became expensive or scarce.

Thomas Mellon only immigrated to the United States in 1819, but he had all the "club" requirements to enter the aristocracy. He was Scots-Irish and a Presbyterian. As a boy, the Mellons settled in the Scots-Irish and German iron-making district of western Pennsylvania. Thomas Mellon the capitalist, banker, and investor moved the coal and iron industry forward in many ways. While making his first money in the law, he became an owner of J. B. Corey and Company in Braddock, Pennsylvania, in 1859. He had actually owned some coal land in the 1840s, but let others work it. Mellon was fascinated with coal and its investment potential. In 1855, Pittsburgh bituminous coal accounted for only eight percent of the pig iron production but its use was doubling every seven years, and by 1880 it would account for the major fuel in pig iron production. Overall use of Pittsburgh bituminous coal was doubling every ten years. Mellon was a true visionary in seeing the future of coal. Mellon also studied the pig iron industry and in 1855 correctly predicted that lake ore would soon replace local low-grade iron ore, allowing the pig iron industry to boom. Mellon became the owner of an arms foundry in Braddock during the Civil War. Maybe the most significant event was his marriage to Sarah Negley. It would be a powerful alliance of the German and Scots-Irish pig iron aristocrats on a par with the marriage of the Craig and Neville families decades earlier. While the Negleys were primarily farmers and rich Pittsburgh landowners, they came from a German family of blacksmiths. Alexander Negley, who came to Pittsburgh after the Revolutionary War, was a blacksmith and military officer. Some of

Thomas Mellon's sons actually considered a career in blacksmithing before taking over the family banking business.

The use of coal mushroomed, but the aggregate replacement of wood as an energy source did not occur until 1880. By 1850, pig iron production had moved from 50,000 tons in 1810 to 600,000 tons in 1850, most of which was from coal energy. The rise in demand in the 1840s opened the door to cheap Irish labor and the beginning of the end of the crafts system in mining. The demand for coal created the necessity of mass production versus crafts production, and mass production meant mass immigration.

The Irish came slowly, taking over the pure labor jobs and supplying children to work in the "breaker" processing plants that sized eastern anthracite coal. Some Germans also came to the eastern mines as well because of the labor shortage in the mines, and a surplus of canal diggers. This is the same template used by new immigrants moving into American growth industries. Their flexibility to move from agriculture to industrial labor defined the nature of immigrant American labor. No doubt the Irish derived some happiness in taking away work from English craftsmen in the coal mining industry. Apprentices performed most of this labor under the British craft system. The transition was rather slow in the 1840s, so as not to alarm the English, Welsh, and Scottish miners. The slow increase of Irish laborers would lead ultimately to unionization and the beginning of the end of mining as a craft. What was different for the Irish in the mining industry was that they were trading up to skilled jobs versus replacing slave and indentured servants in canal digging. In the mining industry, the rise of the Irish was more like their rise in the governmental hierarchy of New York. They were willing to start at the bottom and take one step at a time. The Irish climb was methodical and persistent; and often, while hated, they were underestimated as a real job threat because of their perceived heavy drinking and carousing. The Irish, on the other hand, used their hatred of the English to fuel their drive. Often their Scots-Irish mine owners had no trouble in selling out the hated English craftsmen for the merely disliked Irish laborer. After the Civil War, the Irish would dominate the coalmines of Pennsylvania.

The coal industry would be the lead industry in a great upward economic spiral and co-dependency with the iron and steel industry. The need of coal on the southern East Coast inspired investment in the Baltimore and Ohio Railroad. The railroad would connect the bituminous coal fields of western Pennsylvania and eastern Ohio. By the 1840s, coal was flowing to Baltimore via the railroad. The Baltimore and Ohio pioneered the switch from wood as locomotive fuel to coal in the 1840s. In 1856, the B&O Railroad was consuming 75,000 tons of coal a year, and by the 1870s, the railroads were the principal consumers of coal. The steel industry arose to manufacture steel rails for the railroads, and that required more coal to fuel the blast furnaces. By the 1890s, the steel industry became the biggest user of coal and the railroads' biggest customer. The coal-railroad-steel industry was creating demand that quickly used up the country's unemployed Irish. The need for mass production allowed for the crafts methods to abandoned by owners for mass production techniques. Owners changed their pay scales from minimum bushels dug to piece rates and lowered the piece rate. Furthermore, the owners moved to a daily wage when possible. The Irish who had been successful in the 1850s in unionization found themselves set back a decade.

The King of the coal industry would be Henry Clay Frick.

Owners such as Henry Clay Frick sent agents to Europe and, in particular, southern and Eastern Europe to recruit workers. The agents offered to pay for the ocean voyage (it would later be deducted from their pay). In the early 1880s, massive numbers of Slovaks and Hungarians were brought in. These immigrants were poor farmers with no mining experience or skills. Frick flooded the mines with cheap labor and ended the craft of coal mining. Now even the Irish were upset with the influx of non-English speaking miners. Placards were common decrying the invasion: "One of the most degrading influences brought to bear upon our community is the indiscriminate portion of Hungarian serfs and their employment in public works in preference to good local citizens who are willing and can perform more and better labor for the same pay." The massive influx of immigrants gave the mine owners the advantage in Pennsylvania and West Virginia. Labor was plentiful and wages would be driven down. Frick's success in the coalmines of the 1880s inspired his steel partner Andrew Carnegie to use Hungarians and Slovaks in the steel mills. The successful flooding of Hungarians and Slovaks was followed up with Italians, Russians, and Croatians throughout the 1890s and early 1900s. The last to arrive after 1902 was a wave of Poles. By 1920, the fifty thousand Polish miners were the largest group in the mines. The Poles moved west to dominate the Illinois coal mines as well. Only the Southern coal mines were unique, being manned by fifty- to sixty percent African-Americans (they represented less than five percent in the northern mines). Frick's use of immigrant labor cut the cost of coal in half.

CHAPTER NINE

The Wheeling Iron Aristocracy

The furnaces along the Arundel ore formation in Virginia and Maryland formed the cradle of the Pig Iron Aristocracy. In the 1750s the Principio Works started to buy out other furnaces along the Arundel formation, and consisted of four furnaces, two forges, 30,000 acres of land, and over a hundred slaves. Principio controlled more than half of the exports by 1760. The furnaces utilized flatboats to take the pig iron down the Susquehanna River to British merchant ships at Baltimore. In 1767, Principio manufactured over 2,500 tons of pig iron, while all of Great Britain produced 17,000 tons. Success brought more colonial capital into the firm. Charles Carroll of Carrollton, Maryland, a signer of the Declaration of Independence, took an interest in the firm. Principio produced cannon and ball to protect the port of Baltimore during the Revolution. While Principio was a regional operation, its main owner, Thomas Russell, ran it as an iron plantation with slaves. In 1837, Pig Iron Aristocrat George Whitaker purchased Principio, and ran it as a true manufacturing company. Whitaker brought in the North Forge of Wheeling, West Virginia (then Virginia).

The Whitaker family would make Principio a long-term supplier of pig iron to the expanding rolling mills of Wheeling in the 1840s and 1850s.

Pittsburgh ironmasters often looked to nearby Wheeling to expand the production and uses of pig iron. Wheeling was located on the Ohio River in the then Virginia Panhandle. It was the starting point of deep-water navigation on the Ohio River. The National Road came through Wheeling in the 1820s, which allowed the iron ore, coal, and limestone to be readily available to Wheeling. Particularly, the rich iron ore of nearby Fayette County, Pennsylvania, was available. It had been none other than Henry Clay who had brought the National Road to Wheeling. Like Pittsburgh, Wheeling had rich seams of coal. Wheeling had a deeper riverbed than Pittsburgh.

The man who brought iron to Wheeling was Pig Iron Aristocrat Dr. Peter Shoenberger, who was a Pittsburgh ironmaster and descendant of the central Pennsylvania iron plantations. In the 1790s, the Shoenbergers were in the iron business in the Juniata region of Pennsylvania. George Shoenberger (Peter's father) founded an iron furnace and forge in the Juniata region of central Pennsylvania. The Shoenberger mansion was in Blair County, Pennsylvania, and both father and son built furnaces in the region. George Shoenberger, a German ironmaster, became a world expert in the forging of pig iron into wrought iron. His "Juniata iron" commanded a premium throughout the world. George and Peter soon saw the advantages of processing pig iron in Pittsburgh and established Juniata Iron Works there in 1824. At the same time the Shoenbergers built more furnaces in central Pennsylvania to supply the rolling operation in Pittsburgh. Shoenberger also built furnaces near Niles, Ohio, and Mercer County, Pennsylvania, to supply his Pittsburgh rolling mill. The Wheeling area offered a logical extension of his pig iron empire.

Peter Shoenberger teamed up with another Juniata ironmaster, David Agnew, to build a complete ironworks in Wheeling in 1832. The reason was the rich coal of the Wheeling area. Like Pittsburgh coal, it made good coke as fuel, but it was also known as "steam coal," particularly efficient at running steam engines. Shoenberger's mill would be totally steam powered to utilize the Wheeling advantage. This state-of-the-art works had a puddling operation, blooming mill, forge, a plate mill, and several nail machines. The mill was known as Wheeling Iron Works or by its local name, "Top

Mill." Shoenberger remained in Pittsburgh, while David Agnew became a Wheeling aristocrat. Shoenberger and Agnew later would tie together interests in Ohio as well. Agnew became the major investor in Wheeling's Manufacturers and Mechanics Bank.

Agnew and all of the nation's bankers and iron manufacturers suffered with the Panic of 1837. The Panic was a result of England's decision to maintain the massive labor force in England by dumping cheap pig iron and products on the American market. At the same time, Andrew Jackson lowered tariffs, allowing English pig iron to flow freely into America. Jackson further demanded the use of gold and silver in the purchase of public land, which created a credit crunch. It was a true economic depression, as bank and factory closings swept across America. It would be the start of the debate on Henry Clay's protective "American System." Agnew, needing cash, sold out to John Shoenberger, whose conservative business methods allowed him to ride out the panic. Agnew returned to his iron plantation roots, operating pig iron furnaces in the tri-state iron triangle. Agnew would be the first to use Lake Superior ore in 1853 at Sharpsville, Pennsylvania. Later he would become an investor in lake ore transportation that would ultimately make Pittsburgh, Cleveland, Wheeling and the Mahoning Valley the pig iron centers of the nation and usher in the steel industry of the 1870s. The passage of Henry Clay's 1842 Tariff would ultimately save the area from destruction. Agnew would even team up with Shoenberger to build furnaces around Sharon, Pennsylvania, and Youngstown, Ohio. Again the political foundations were set for the area to become a stronghold of the Whig Party and its protectionist plank.

The protective tariff of 1842 stimulated more investment from the Pig Iron Aristocrats of Pittsburgh in the Wheeling district. One of these Pittsburgh ironmasters was the shadowy E. W. Stephens. Stephens had been in the nail making business in Pittsburgh since 1830, and he was a key political supporter of Henry Clay. Little more is known of Stephens except he had plenty of cash to lease the Top Mill from Dr. Shoenberger. Stephens brought with him some extremely talented ironworkers such as such as William Bailey and the Norton brothers: Edward M., Fred D., and George W. The Nortons revolutionized the nail industry. In particular, Edward Norton was an extraordinary mechanical engineer and manager. In 1847, Edward Norton split off, building the Virginia Mill of the newly formed Norton, Bailey and Company. The Virginia Mill was state-of-the-art, and quickly made the "Wheeling nail" famous. The Clay tariff created an iron business boom in the nation, and Edward Norton moved on to build the Belmont Iron Works in Wheeling in 1850. Belmont Iron could produce two hundred kegs of nails a day. It had not only six nail machines, but six puddling furnaces and two rolling mills. Within another year, the same investors created La Belle Mill of Bailey, Woodward and Company. The La Belle Mill was even larger than Belmont, with eight nail machines and eight puddling furnaces. The four nail mills were controlled by an interlocking directorship that still included Shoenberger, Stephens, and Bailey (this was prior to anti-trust legislation). Wheeling nail production was overtaking that of Pittsburgh and the competition for pig iron in the market was driving prices up.

Wheeling's real appearance on the pig iron manufacturing scene came with the building of the massive Crescent Iron Works on Wheeling Creek in 1854. It hoped to take advantage of Wheeling's coal and superior transportation network. This monster works had fifteen double puddling furnaces. Edward W. Stephens financed this mill behind the scenes. While this mill was designed to make various products such as sheet and boilerplate, its new feature was the ability to roll iron railroad rails. The railroad rail market was a British monopoly at the time, but Crescent believed it would be able to undersell the British at a considerable profit. The plan was also to build a large pig iron blast furnace. The Crescent Iron Works would bring Maryland and Pennsylvania ironmaster Joseph Whitaker to Wheeling through the Principio Company that became a major supplier of pig iron for the Crescent Mill. The Whitaker family, being from the old iron plantation masters, also brought political power to Wheeling. George Whitaker was a member of the Maryland state legislature and a director of the Baltimore and Ohio railroad. The great works was built but several factors caused it to close within a few years: primarily, the Panic of 1857 and the inability of Wheeling Creek to support barge traffic. It would reopen with the start of the Civil War under a new management group with many of the original investors such as Chester Hubbard. The Whitaker family would become the most representative of the Pig Iron Aristocrats. The Whitaker family also remained invested. George's son, Nelson Whitaker, became a U.S. Senator in the 1890s, and remained invested in the Wheeling iron industry. The Principio Company supplied pig iron into the late 1890s. Its properties would eventually pass to the Wheeling Steel Company and then to Pittsburgh Wheeling Steel.

The problem of obtaining cheap pig iron had been growing as western Pennsylvania's charcoal furnaces and those of the Mahoning Valley were running out of resources such as hardwood and iron ore. The use of coal had improved things temporarily. The Hocking Hills furnaces were shipping pig iron up the Ohio River to Pittsburgh and Wheeling. In the late 1840s, two large iron ore deposits in Missouri had given birth to a pig iron industry. Missouri pig iron also flowed up the Ohio River. The Wheeling Pig Iron Aristocracy built the Missouri Iron Works in Wheeling to expand nail manufacture further, using Missouri pig iron. Pig iron production was big business now key to Pennsylvania, Ohio, Virginia (West Virginia), Maryland, Michigan and Missouri. The finding of high grade ore brought in Pig Iron Aristocrat investment and control. Missouri ore could readily flow up the Ohio to Wheeling. Missouri lacked the coal being used in the latest blast furnace technology. In 1857, across the river from Wheeling at Martin's Ferry, Wheeling's first blast furnace was built.

Cyrus Mendenhall was a Western Reserve Quaker. He originally had discovered iron ore on his farm about four miles from Wheeling, which he hoped to use to produce pig iron for the booming Wheeling nail mills. He would soon switch to richer ores. Mendenhall was a trader who would bring Missouri ore and Lake ore (and even some Spanish ore up from New Orleans) to operate his blast furnace at Martin's Ferry. Mendenhall employed Youngstown ironmaster John Maycock to operate his blast furnace.

Maycock, using local soft coal, fired the furnace and produced about ten tons per day. He would eventually use Connellsville coke from western Pennsylvania which could readily be brought down the river.

CHAPTER TEN

The Rise of the Blast Furnace

Until the 1850s, most pig iron was produced at the iron ore districts. Western Pennsylvania, Maryland, Virginia, eastern and southern Ohio, Missouri, and Michigan were the major pig iron producers. Pittsburgh, Youngstown, Wheeling, Cincinnati, and Cleveland became the major consuming centers. In addition, prior to 1850, most furnaces were fueled by charcoal, and hardwood was becoming scarce. Even with favorable tariffs, England supplied a sufficient amount of pig iron and iron products. England supplied most of the railroad iron rails until the late 1860s. British iron was also considered of higher quality. The American iron producers were using eighteenth century British technology. In 1850, British furnaces were larger and were using coke as a fuel. America had suffered from its own abundance of hardwood. Charcoal was cheaper, but it limited the height of the furnace stack. But things were beginning to change by the 1850s.

The demand for pig iron by Pittsburgh foundries and rolling mills was huge. In 1857, Pittsburgh had twenty-five rolling mills producing nails, sheet iron and plates of iron. The rolling mill demand was around 150,000 tons of pig iron a year. The foundries, which re-melted pig iron, had grown to sixteen with thirty associated re-melt furnaces using 50,000 or more tons of pig iron a year. There were also related heavy industries such as steam engine manufacture and forge shops. Yet, no railroad rails were being made, which appeared to be the future for pig iron production, but the industry would need to change to challenge the premier pig iron rails of Britain.

That change would come from a variety of sources. First the political realization of the need for America to be self-sufficient in ironmaking, and the evolution of the Republican Party to support that view. The discovery of rich ores in Michigan and Minnesota ended the rich ore shortage. Lumbering of hardwood created a shortage and movement towards America's abundant coal deposits.

The Development of Large Blast Furnaces

The advance of steelmaking technology would open new markets for iron. The Civil War would also create demand for iron products as never before. And maybe most significant was the growth of America's railroads, which became the major user of iron products. The railroads also created the transportation network to bring it all together. Railroads became the largest consumer and supplier to the iron industry. American iron rail producers would have to move to coke technology to assure the needed quality to compete with British iron rails.

The great ironmaster and iron plantation manager, Isaac Meason, seemed to have pushed the use of coke in his Fayette furnaces first in America. Meason was aware of the coal technology of Scotland, and saw its future in a number of industries. Meason's charcoal furnaces were stripping an acre of woodland a day (800 bushels of charcoal). Charcoal was extremely labor intensive with dozens of woodchoppers and around 12 colliers assigned to a single furnace. He appears to have had hired English ironmasters in 1800 to help him make the conversion. His Fayette furnaces were ideally located on the Connellsville bituminous deposits of coal. By 1812, Meason had built coke ovens and was using coke throughout the operation. These earlier plantation blast furnaces could easily double or triple daily production of the charcoal furnace. Meason's coke-fueled furnaces were some of the most profitable, and exceeded the British quality. His castings such as stoves, sugar-making, and salt-making kettles went all over the country. Meason's

The Rise of the Blast Furnace 121

pig iron also dominated the Pittsburgh market in price and quality. His technology would put him in the lead over western Pennsylvania iron furnaces, and made him a fortune.

Unfortunately, Meason was far ahead of his times in the 1810s. By 1840, some Pennsylvanian Scots-Irish furnace managers were experimenting with anthracite coal and hot air blasts. This technology was equivalent to coke, but limited to eastern furnaces where anthracite was abundant. Coke iron furnaces would not overtake charcoal furnaces until the 1870s. The real breakthrough with big production coke blast furnaces came in 1859 with the building of the Clinton furnace in Pittsburgh. Graff, Bennett, and Company built the Clinton Furnace on Pittsburgh's south side. The Clinton furnace used coke that had been made from Pittsburgh coal with very poor results. Experimenting with Connellsville coke turned things around. The experiment demonstrated what the country would soon agree on: Connellsville coke was unique and the best blast furnace coke in the world. This experiment would bring a rush of Pittsburgh investors into building blast furnaces, and make Henry Clay Frick one of the world's richest men. Connellsville coke would also be used at the Wheeling blast furnace of Cyrus Mendenhall. The iron ore came from the Missouri and Lake Superior Region via the Ohio River, and in the case of the lake ores, by canal. Another key factor in the development of the blast furnace was the introduction of Lake Superior ores. Ore was shipped from the lake ports of Cleveland and Erie, and then moved to the Youngstown area.

The rise of the modern blast furnace in the 1850s was truly evolutionary in nature. Coke and better ores was one part of the equation. Hot air blasts were also critical. Charcoal furnaces had used water-driven bellows to inject air into the furnace for better combustion, but cold air also removed energy because it had to be warmed in the furnace. In the 1840s, several creative Scots-Irish plantation owners started to experiment with hot air blasts. Pipes on top of the furnace that cycled hot air to the bellows achieved this. Eliza Furnace in Cambria County, Pennsylvania had perfected this by 1848. Also in the 1840s, steam engines were used to replace water-driven bellows, which significantly improved the efficiency and productivity of the blast furnace. Another improvement was water-cooled tuyeres. Tuyeres were nozzles in the furnace that injected the air blast. The tuyeres were water-cooled copper that would melt in the furnace without water cooling. The use of coke, steam engines, and hot blast eliminated most of the geographic constraints of the charcoal furnaces. The old iron plantations were located where both fuel and iron ore was abundant; now, transportation equalized the playing field. Still, coal districts were favored locations because iron ore was cheaper to transport. Iron ore was heavy and dense compared to coal or coke.

Many look to the Buena Vista charcoal furnace of Connecticut as the first "blast furnace." Built in 1847, its ten-foot bosh and twenty-nine-foot stack made it the largest charcoal furnace at the time. It was water-powered but used hot blast technology. While Buena Vista had a forge, it was primarily a pig iron producer. It did have a complex and improved tuyeres system not usually found on charcoal furnaces. It used the rich hematite ore of Salisbury. Buena Vista supplied pig iron to New York, Delaware, and New Jersey foundries and iron works. It was also supplying the Ames family shovel plant down river. The charcoal supply of New England appeared more limitless at the time. Hardwood would ultimately be the major limitation of the charcoal furnace, but it would be 1890 until coke furnaces overtook charcoal furnaces, and Buena Vista operated until 1894.

Another part of the limitation on charcoal furnace production was poor construction. Cut stone, often limestone, was used to build charcoal furnaces for over two hundred years. Commonly the cost was near $50,000 in the first half of the nineteenth century. A furnace would take weeks to start up and would run for nine months before re-lining was needed. As furnaces grew in size, fireclay bricks were used to extend the life of the furnace lining to years. The use of coke also reduced the start-up time for a new lining. Furnace investment cost did however, increase dramatically with better construction and linings. These higher costs changed the nature of pig iron production from "mom and pop" operations to corporate investments.

The father of the modern blast furnace was Pig Iron Aristocrat James Ward of Niles. His experiments and knowledge made him the leading authority on furnace operations. In the 1840s and 1850s, Ward operated a number of blast furnaces in the Youngstown area. His Falcon Furnace was considered the most productive in the world in 1854. In 1854 Joseph Brown and his brothers, of old Pennsylvania iron aristocracy, purchased the small operation in Youngstown and expanded it. Brown added puddling furnaces and purchased Falcon Furnace to feed the operation. Brown also added another blast furnace called Phoenix Furnace. These furnaces supplied Youngstown rolling mills and sold pig iron on the open market. Falcon Furnace supplied the enormous appetite of the Pittsburgh rolling mills for pig iron. In 1859, James Ward built his monster Elizabeth Furnace, which stood sixty-five feet (the highest at the time) and had a fifteen-foot bosh (also the largest at the time). This furnace achieved forty tons per day, making it the most productive in the world. It used Lake Superior ore, hot blast, and coke. This furnace would run into the 1940s as the Hanna Furnace of Republic Steel. Ward's technology produced a number of competing furnaces in the Youngstown area, including Phoenix, Nimrod, and Grace Furnace. Youngstown was becoming the heart of Pig Iron Aristocracy because of its canal connection for lake ore and its coal deposits. Youngstown's canal also offered a route for pig iron to flow to the rolling mills of Pittsburgh. Youngstown district was the center of pig iron production during the 1850s.

The early blast furnaces of the 1850s were built to produce pig iron for other iron operations. The blast furnace could stand on it own, supplying many different foundries and rolling mills at points distant from the operation. The pig iron producers started to look at more integrated operations during and after the Civil War. The story of the next blast furnace in Pittsburgh is the story of Jones and Laughlin Steel. That story begins with the world's first oil baron, Samuel Kier and Keir's young bank manager Benjamin Franklin Jones. Both men had invested earlier in canals. In 1850, Kier focused his energy on oil, while Jones invested in puddling furnaces and rolling mills on Pittsburgh's Southside (known as Brownstown). Another Pittsburgh iron investor, William Thaw, teamed up with Kier to extend canals to connect Pittsburgh and Youngstown. Benjamin Jones was Scots-Irish from a modest background, but as a young boy showed an aptitude for business. Jones had been manager of the Pennsylvania Canal prior to this move in 1850, but in the 1840s he was a partner of a charcoal furnace and forge in Pennsylvania's Indiana County. The furnace failed with a change in national tariffs, allowing more imports. Jones had got the bug for the pig iron industry and its potential. The depression of the pig iron industry caused by lower tariffs left a lasting impression. Jones switched from the Democrat Party of his upbringing to the protectionist Republican Party. B. F. Jones would become the nation's leader in support of a protectionist tariff policy. He would become an important member of the old First Presbyterian Church of earlier Pig Iron Aristocrats.

Pig Iron Aristocrat Jones never got his Southside works running, but in 1853 he teamed up with two German ironmasters, Bernard and John Lauth. Samuel Kier was also an investor in Jones, Lauth, and Company. The 1853 partnership agreement of B. F. Jones would augur that of the future Steel King, Andrew Carnegie. The agreement called for all profits to be re-invested in the operation. Furthermore, the partnership was the first to apply the idea of vertical integration in the pig iron industry. Vertical integration meant owning the full manufacturing chain from raw materials to distribution. This was the idea of Jones, who opened warehouses in Chicago and Philadelphia to market iron products. Jones focused in particular on the needs of the railroad industry. The operation in 1855 was known as the American Iron Works, and got its pig iron form the Falcon Furnaces of Youngstown, Ohio. The plant rolled bar and rails. The Lauth brothers were brilliant engineers and the American Iron Works became the most productive rolling mills in the country. The American Iron Works soon became a major consumer of pig iron. The Lauths also invented the cold rolling process to produce polished bar surfaces.

The Lauths were German-Catholic and could never gain acceptance in the elite Pig Iron Aristocracy of Pittsburgh. Their contributions were critical to the success of the operation, but they took the opportunity to sell their interests in 1861.

B. F. Jones took vertical integration to new heights, while some Youngstown pig ironmakers had combined coal, coke, and ironmaking in late 1859. Jones partnered with Connellsville coal operations and river transporters to bring coal to his beehive coke furnaces. He owned several of the Pennsylvania coalmines such Vesta Mines. The coke furnaces fed his Eliza Furnaces at his American Iron Works. He purchased iron ore mines in Michigan and Minnesota, investing in canals and railroads to transport ore. The works had puddling furnaces and rolling mills to produce iron products. Benjamin Jones and his brother took to "forward" integration as well. They built warehouses in Chicago and Pittsburgh, initiating the idea of iron and steel processing centers for the end-user.

In 1859 another partner, Scots-Irish James Laughlin, a Pittsburgh banker, came into the partnership. Earlier James Laughlin's Youngstown furnaces were supplying the American Iron Works of the Lauth brothers. The pig iron requirements for the rolling mills were soon outstripping the ability of Pennsylvania and Ohio furnaces to supply them. As part of Jones's vertical integration he set up two blast furnaces (Eliza Furnaces) on the north side of the Monongahela River. Naming blast furnaces after women was a tradition. These furnaces were forty-five feet high and had a twelve-foot bosh. Additional integration included the building of coke beehive furnaces. Coal was brought down the river from Connellsville. Iron ore was brought in from Missouri and the Great Lakes. Vertical integration allowed Jones and Laughlin to become the low-cost producer, a lesson that a young Andrew Carnegie would note. The basic layout of the operations would remain intact until 1980! Thus in 1861, the Jones and Laughlin Company became the first integrated iron works. Both Jones and Laughlin worked behind the scenes with the Pig Iron lobby to assure Lincoln's protectionist policies. Their support was a key reason behind Lincoln carrying Allegheny County.

One difference between banking control and the Pig Iron Aristocrats was who controlled the operations. The Pig Iron Aristocrats were operators first. They used their banking connections and political connections to finance and invest in expanding their operations. They loved the size and mastery of their furnaces and mills. They rarely became politicians, looking at politicians as lower class. When the corporate steel kings took over in the 1900s, the bankers took control of the operations, and the politicians limited operations. The Pig Iron Aristocrats expanded into older iron districts with the political success of the protectionist Republicans.

Like Jones and Laughlin, who hoped to supply the iron rail market, Cambria Iron Works opened in 1853 in Johnstown, Pennsylvania. The Cambria furnaces could produce over 150 tons of pig iron daily. The Johnstown area also was blessed with bituminous coal and iron ore. Cambria rolled its first wrought iron rails, but had problems penetrating a British monopoly. This would start the political push to protect the infant rail industry of the United States. The Pig Iron Aristocracy was able to win its plank for protectionism in the newly formed Republican Party. Still, the American rail industry had to prove itself on quality. Thomas Scott and Edgar Thomson of the Pennsylvania Railroad pushed for rail trials in a hope that American production could reduce track costs. Ultimately, the American railroads and the Pig Iron Aristocracy would form a symbiotic alliance that would change politics in America.

Protectionism allowed the Pig Iron Aristocrats to invest for the future because of a stable market. Prior to the political movement towards aggressive protectionism, the pig iron market had large economic swings. There was a cycle of boom and bust, which favored running out furnaces and equipment, versus long-term investment. Technology was of no value if pig iron producers were afraid to invest. England was known for its policy of dumping to prop up its pig iron industry. The pig iron industry meant high employment, and countries such as Britain realized its value. Until the 1850s, the Pig Iron Aristocrats had had mixed political results in iron tariffs. The rise of the Whig and Republican Parties united the industry, which had been fractured by Democratic Jacksonians. The South, in particular, was represented by free trade Democrats. The division on iron protection like slavery was becoming a north-south split.

Lincoln's future economic advisor, Henry C. Carey, was a huge supporter of Clay's American System of protection. Carey's writings became more popular in the North than those of Adam Smith. Carey saw the free trade of the British as a great error, stating in his 1851 book, *The Harmony of Interests*, that free trade would lead to the fall of industry. Democratic control of the White House in the mid-1850s only increased the resolve of the Republicans and the political support of the Pig Iron Aristocracy. Tariffs were reduced under Democratic administrations from Jackson on.

Lincoln's presidential victory of 1860 was due to his success in the old "Iron Whig" districts of Ohio and the other Pig Iron Aristocracy districts in Pennsylvania, Maryland, Connecticut, and western Virginia. The vote in these districts set new majority records as Lincoln's protectionism played as well as his anti-slavery stand. The election would bring war to the nation, but prosperity to the iron districts of the North. On his way to Washington, Lincoln would stay over at Pittsburgh's Monongahela House, where he made a speech to more than 10,000 people. Congress passed the highest iron tariffs ever, along with an increase in tariffs across the board. At the time, the government's main source of income was tariffs (not income taxes). The protectionist representatives wrote the tariff bill, assuring iron received the highest level.

Pittsburgh, West Virginia, and eastern Ohio represented the heart of Republican support as the forge and arsenal of the North. In 1861, Pittsburgh had over three hundred puddling furnaces, three of the largest blast furnaces in the country, twenty-six foundries, twenty-five rolling mills, seven chain factories, six forges, sixteen machine shops, and one hundred related factories. Within a hundred miles in the Mahoning and Hocking valleys and hills of western Pennsylvania were hundreds more furnaces, rolling mills, and metal factories. The Hocking Hills furnaces of southern Ohio were considered producers of the best pig iron for cannon in the world. Wheeling had seven large rolling mills, seven foundries, and three nail mills. This was the strategic triangle for manufacturing, and it was close enough to the South to be threatened.

During the Civil War, the 23rd Ohio regiment had been assigned to protect the iron districts of Ohio and Pennsylvania. This regiment was commanded by future president, Rutherford B. Hayes, and another future president, William McKinley, was an officer in the regiment. In October 1862, the 23rd was ordered to Pennsylvania to protect against possible cavalry raids on the iron valleys of Pittsburgh, and possibly even Ohio's Mahoning Valley. Through the spring of 1863, there were rumors of Confederate cavalry near these manufacturing centers. Most of these never proved out, but Hayes did cut off an attempted raid into Ohio by Jim Morgan. The southeastern counties of Ohio were the major producers of pig iron during the war, and were targets for the Confederates. This "Hanging Rock Iron District" was a thirty-mile belt in the counties of Scioto, Lawrence, and Gallia on the Kentucky border. Some of the largest Union blast furnaces were in this strategic area. This district had over fifty charcoal furnaces, supplying rolling mills and the foundries of Pittsburgh. The area also had foundry operations, and the Hecla Furnace produced the famous Swamp Angel cannon used in the siege of Charleston Harbor. The high quality of Hocking Hills pig iron made it the preferred raw material for iron cladding of ships such as the *Monitor*. Even the British had found this high-quality Ohio ore superior to their own, and had used large amounts of it to produce armaments during the Crimean War (1854-1856). The 23rd was one of many regiments used to defend against planned Confederate raids. This area would be part of the "Pig Iron Aristocracy" that supported McKinley's and Hayes's presidential runs.

One of the battles to cut off the Confederate Morgan in the Hanging Rock area was the battle of Buffington's Island. The

The Rise of the Blast Furnace

Confederate general had entered Ohio near Gallipolis about thirty miles from some of the Union's largest blast furnaces in the summer of 1863. The 23rd Ohio, the 13th West Virginia, Union gunboats, and Union regulars amassed to chase off Morgan's cavalry. At Buffington Island this combined force killed over eight hundred and captured 2,300 Confederate raiders (208 by the Ohio 23rd).[14] Still, over three hundred of Morgan's cavalry moved north towards the Mahoning Valley area. Every able-bodied man in the valley took up arms and marched south. Panic struck the area, with banks moving money to Cleveland. Finally Morgan was forced to surrender at West Point, Columbiana County, about twenty-five miles from Youngstown (it was the farthest point north reached by the Confederate Army in Ohio). Pittsburgh had many additional fears during the Battle of Gettysburg, when Confederate cavalry was headed for Pittsburgh. The war did, however, create the greatest demand ever for pig iron, and the protectionist policies of Lincoln and the Republicans assured that it would translate into the growth and expansion of the American pig iron industry.

---

[14] William H. Armstrong, *Major McKinley* (Kent: Kent State University Press, 2000), 55.

CHAPTER ELEVEN

War Demand

The Pig Iron Aristocrats of Ohio and Pennsylvania had brought war, not by their abolitionism, but from their protectionism. The iron districts of Ohio, Maryland, Connecticut, and Pennsylvania had given Lincoln a resounding victory in 1860. Lincoln had been the first national candidate since Henry Clay to have the united support of the Pig Iron Aristocrats. Lincoln carried Pittsburgh's Allegheny County by a record 10,000 votes. Lincoln called the concentration of votes in this area "the State of Allegheny." Pittsburgh had finally overcome the divides of the 1794 Whiskey Rebellion. The margins were similar in the iron districts of Ohio. The Iron Whigs and protectionist Democrats had found a new home in the Republican Party. In western Virginia, the Pig Iron Aristocrats' support of the protectionist Lincoln split the state and created West Virginia. These Pig Iron Aristocrats had forged an alliance with iron labor as well. A strong pig iron industry was necessary for both management and labor. The German and Irish immigrants of the 1840s came for economic opportunity and they united with wealthy Scots-Irish to form a new Republican machine in the iron districts. Industry growth took priority over unionism and profits. These districts knew the recessions of free trade policies. As a result of war and protectionism, the pig iron industry would see great advances in technology. The Pig Iron Aristocrats were rewarded for their votes with the 1862 tariff act, which was the highest ever on pig iron at thirty-two percent. As the

Pig Iron Aristocrats responded with massive investments in industry, the Congress moved the rate to forty-seven percent in 1864. The pig iron industry grew an amazing sixty-five percent during the Civil War. By the end of the war, the Pig Iron Aristocrats were a real national political force with wealth and the ability to employ tens of thousands.

Besides technology, demand would drive pig iron production, and there is no demand such as that of war. As much as twenty-five percent of the Union's artillery (fifteen percent at Fort Pitt Foundry alone) was made in Pittsburgh. At least eighty percent of the Union's naval iron plate for ships and most of the Union's armor plate was rolled in Pittsburgh. All of the artillery carriage axles and most railroad axles were forged in Pittsburgh. But most of the raw pig iron, however, came from Ohio. The Pig Iron Aristocrats were not only the ones who won the war, but the ones who profited the most. The Republican tariffs assured a boom in national production. The great iron triangle of Ohio, West Virginia, and Pennsylvania saw growth as never before. The war would also stimulate huge leaps in pig iron technology. The huge profit margins in the pig iron related businesses assured those profits were poured back into the businesses. Important pig iron end users such as the railroads experienced similar growth. The expansion of American industry during the war would put the infrastructure in place to make America the premier industrial nation.

Another part of the pig iron industry often overlooked was that of lower Connecticut and New York. Charcoal furnaces like Buena Vista were supplying a number of armories and foundries. The Ames Ironworks had moved from shovels to cannon in the late 1830s, and in the 1840s, it cast the largest cannon to date, a huge Parrott gun. At the nearby Cold Springs armory, Parrott guns were being mass-produced in preparation for war. All of the pig iron came from Connecticut charcoal furnaces. The Parrott, with its steel reinforced chamber, was the most popular medium-size gun during the war. The owners invested heavily in cannon production, and the reduction of demand after the war sent it into bankruptcy. Politically, the Connecticut Valley aligned itself with the iron triangle of Ohio, West Virginia, and Pennsylvania. The Connecticut Valley bridged the old Ohio Western Reserve with the New England states. All of these states of the Whig Party of protectionist Henry Clay would form the infrastructure of the new Republican Party.

Pittsburgh's foundries and rolling mills were dependent on the blast furnaces of Ohio's Hocking Hills, western Pennsylvania's Juniata region, and Ohio's Youngstown district for pig iron. The Jones and Laughlin furnaces expanded, but could only supply their south side rolling mills. The war brought not only expansion of the Jones and Laughlin Eliza Furnaces, but also the twin monsters of Shoenberger, Blair, and Company known as the Superior Furnaces. These furnaces, built in 1865, stood sixty-five feet high (the largest in Pittsburgh) and had a dual capacity of 48,000 tons annually. The eyes of the nation were not focused on blast furnaces, but the great foundries such as West Point Foundry in New York, Algers Foundry in Boston, and Ames Foundry in Massachusetts. The greatest producer of mammoth cannon was the Fort Pitt Foundry in Pittsburgh. Fort Pitt Foundry was the direct descendent of the Joseph McClurg works of 1804. This foundry would produce sixty percent of the Union's heavy artillery and fifteen percent of all Union artillery. The Fort Pitt Foundry became famous throughout the war across Europe, as reporters marveled at the huge guns cast there.

Fort Pitt Foundry had led the country in the development of heavy cast iron in the 1850s. Fort Pitt Foundry in the 1850s was located on the north side of Pittsburgh (Allegheny City) on 28th Street. Charles Knap then owned the foundry, having purchased it from the sons of Joseph McClurg in 1841. The real advance in heavy artillery came from the work of Army Major Thomas Jefferson Rodman. Major Rodman was the superintendent of the Watertown Arsenal in Massachusetts. Large cast cannon had become problematic, since these hot cast cannon cooled, creating internal strains. The strains would cause the cast cannon to break on cooling, split during transport, or burst in firing. Rodman had worked out a revolutionary process of cooling the cannon from the inside core. This allowed for a hollow tube to be cast. Prior to this, large cannon in America and at Germany's Krupp Works were cast solid and bored out. Rodman had received a patent in 1847, but the Army was not impressed. In 1849, Rodman signed a contact to develop large cannon at Knap's Fort Pitt Foundry.

Rodman moved to Pittsburgh and started a series of production experiments that made news around the world. Over the next ten years Rodman perfected his process of hollow cast cannon. The larger size of the Rodman cannon posed additional technical problems beyond casting. Pig iron purity was critical to prevent bursting, and Rodman settled on Hocking Hills pig iron because of it purity. Pig iron was then re-melted in improved air furnaces. The breech was another area of concern for larger cannon. Solid cast cannon had moved to the use of a wrought iron hoop around the breech. Robert Parrott of West Point Foundry in New York had developed this technique. John Dahlgren had cast bottle shaped breeches to better withstand the greater gunpowder charge. Krupp Works in Germany was experimenting with tougher steel castings. Rodman did increase the width of the breech like Dahlgren, but less was needed because of his strain-free casting system and pure pig iron. In 1860 the Fort Pitt Foundry cast a pair of fifteen-inch Rodman "Columbiads," the largest cannons in the world. The Pig Iron Aristocrats were now the pride of the country and heroes worldwide. The foundry had stockpiled a smaller ten-inch version to supply the Army.

Pittsburgh was truly the forge of the Union. Pittsburgh had over four hundred puddling furnaces in operation during the war. Its rolling mills had an annual capacity of over 150,000 tons annually. Pittsburgh's foundries had an annual capacity of over 34,000 tons. Most of the pig iron was coming in from Ohio to run Pittsburgh's furnaces and rolling mills. The war united the Pig Iron and Pig Iron-related states of Ohio, West Virginia, Pennsylvania, Michigan, Kentucky, Missouri, Illinois, New York, and Connecticut. All of these states would be the iron heart of the new Republican Party. At the same time, Pittsburgh was experimenting with a new iron product known as cast steel. Steel was being made by the "German method," a "high" volume variation of crucible steel. This required pig iron to be heated in crucibles in coal-fired furnaces for two days. Then carbon was re-added to produce steel. One crucible might only hold two hundred to one thousand pounds of steel. Larger castings required a simultaneous pouring and mixing of the liquid steel. Several Pittsburgh foundries such as Singer, Nimick and Company were casting small rifle (three-inch) steel cannons to compete with German and British guns. By 1864, Pittsburgh's Hussey, Wells and Company, using the crucible and German method, could produce twenty tons of cast steel a day (when most steel companies were lucky to cast a ton a day). Steel would never be a factor during the Civil War, but would bring the Pig Iron Aristocrats much wealth, and ultimately, steel would spell the end of the Pig Iron Aristocracy.

The Pig Iron Aristocrats reflected the strong Union and Republican support of the Ohio-Pennsylvania iron district during the war. Just prior to the war, the Allegheny Arsenal in Pittsburgh had a stockpile of large Rodmans. In December of 1860, prior to Lincoln taking office, Secretary of Defense John Floyd offered these cannon to be shipped to New Orleans. Major John Symington, commander of the arsenal and Southern sympathizer, prepared to ship. Word got out that these great guns might fall into the hands of the South. Angry crowds gathered at the Monongahela wharf where the cannon were to be loaded on steamboats. The mob set up a gun battery on Brunot's Island a few miles from Pittsburgh on the Ohio River with the plan to sink the steamboats. Pig Iron politicians telegraphed Congress and the President and got the orders rescinded. Two of these 10-inch Rodmans were purchased by George Westinghouse in 1904 and today can be seen outside Pittsburgh's Soldiers and Sailors Hall.

As the war progressed, Rodman began work on a twenty-inch cannon, which would weigh over fifty tons and fire a half-ton round of shot, up to four miles. The estimated cost of the finished gun was over $32,000. It was the birth of the industrial-military complex that only added to the political power of the Pig Iron Aristocrats. Rodman stated: "it is not deemed probable that any naval structure, proof against that caliber, will soon if ever be built." The casting of the proposed gun was planned for early 1864, and it gained the attention of the world. Reporters and military observers from all over the world descended on Pittsburgh and the supply furnaces in Ohio. There were military officers from Prussia, England, France, Austria, Russia, Spain, Sweden, and Denmark. For this project, charcoal pig iron from Pennsylvania's Juniata district and Ohio Hocking Hills was planned. The mold would require eighty tons of re-melted iron. Because the largest reverberatory air furnace could handle less than forty tons, five furnaces were to be used and tapped simultaneously (another technique pioneered at the Krupp Works in Germany). The core would be water-cooled using the Rodman process.

On February 11, 1864, the world press gathered for the great event. The cast cannon took ten days to completely cool. The gun required twenty-four horses to move it to the Pennsylvania Railroad special car. The weight restricted the train to a speed of thirty miles per hour on its trip to Fort Hamilton, New York. The firing of the gun showed a range of five miles. It was never used in the Civil War, but remained on seacoast duty through World War I. The story resulted in the novel, *From the Earth to the Moon*, by Jules Verne. A few months later, the Fort Pitt Foundry cast another great twenty-inch gun, the XX-Dahlgren known as "Beelzebub." This would be the first of a series of XX-Dahlgrens: Satan, Lucifer, and Moloch. These guns made too late for service in the Civil War. In addition, the foundry made 645 ten-inch Rodmans, 90 thirteen-inch mortars, and 80 fifteen-inch Rodmans. The mass demand on pig iron caused a 100% increase in price and a world-class metal industry. The reports of the manufacture of these great guns in *Scientific American* and *Harpers* put metalworking into the public attention. There evolved a type of national pride in what industry could accomplish. This pride was an important element in forging an alliance between labor and the Pig Iron Aristocrats.

Armor plate production was just as important. Armor plate was produced in puddling furnaces and then rolled into wrought iron plate. The firm of Lewis, Oliver, and Phillips produced most of the Union armor at the Southside and Northside operations. Another armor plate operation was that of Lyon and Shorb, who produced naval armor for ironclad vessels. Most of the armor plate mills specified charcoal pig iron from Hocking Hills. There were two proving grounds in the Pittsburgh area to test cannon and armor. Pittsburgh had at least three foundries producing cannonballs, but often cannonballs were cast directly at pig iron furnaces in Ohio. Wheeling was mainly producing nails for the war effort, but Belmont Mill was producing large quantities of one-inch armor plate. Belmont was the old Shoenberger mill now owned by the Norton brothers. The Hanging Rock district made most of the cannon projectiles with Pittsburgh adding another ten percent. The iron districts not only supplied the cannon and armor, but in the hotly contested 1864 election held strong, saving the election for Lincoln and holding the Union together. The pig iron industry, the Republican Party, antislavery groups, and protectionists were welded together in the 1860s. The Pig Iron Aristocrats had returned to the old power days of James O'Hara in the 1790s. Political decisions were once again made in clubs such as the Duquesne Club, the Presbyterian Church, and even the Masonic lodges. No formal trust was needed; the informal infrastructure was just as unified and powerful.

The demand of the Civil War changed the nature of the iron industry. In 1862, *Scientific American* reported, "The pocket iron deposits of Ohio, West Virginia, and Pennsylvania had either been exhausted or, with few exceptions, were too scattered to feed increasing number of blast furnaces." The problem was critical for the independent blast furnaces of Pittsburgh, Wheeling, and Youngstown, which purchased ore from afar. Wheeling and Pittsburgh turned to Missouri iron ore, while Lake ore continued to be developed and pioneered by Youngstown companies. Wheeling ironmasters Andrew and Alexander Glass played prominently in transporting ore on the Ohio River from Missouri. The war, however, pushed development in Michigan and Minnesota to supply blast furnaces more ore.

A young Union officer saw firsthand the development of a new iron product during the Civil War, which by the end of the century would create a new huge market for pig iron. William McKinley was a commissary officer in charge of supplying an army division. Limited canning capacity and tinplate production in the United States limited supplies to the Union Army. But even the Union was importing ninety-nine percent of tinplate needed. Hussey Company in Pittsburgh was making some trial lots. McKinley took to the clerical as well as the logistical requirements of the work. Decades later President McKinley would impose a tariff that created the American tinplate industry; McKinley's home state, Ohio, would take leadership in its production. Supply management had become a logistical necessity with the invention of the tin can. In the early 1800s, Napoleon offered a reward for a better method of preserving food for his army. Nicholas Appert won the reward in 1810, and contributed to freeing his army from daily foraging for food. The United States lacked the tinplate, being totally dependent on England, and the tinsmiths to keep up with the demand of war. Still, the Union was in better shape than the Confederates, who were dependent on foraging for food. Tinplate and cans kept the Union Army well fed and made a contribution to victory, not overlooked by politicians.

   The Civil War created a new breed of Pig Iron Aristocrats, but like their iron plantation ancestors, they lived in mansions and were bigger than life. But unlike their 1700s ancestors they didn't just go into politics, they controlled politics. They were the core of the new Republican Party, and controlled Pennsylvania, West Virginia, and Ohio. Many of these Pig Iron Aristocrats were investors in the great furnaces and mills such as Henry Buhl and William Thaw (son of John Thaw). The Civil War had brought this political alliance into being. The Pig Iron Aristocrats' drive to grow their industry actually resulted in an end of slavery. They would dominate the party, Congress, the Ways and Means Committee, and the presidency until the 1920s. They would create the new generation of steelmakers in the 1880s. The massive labor requirements of the pig iron and related industries would create the great wave of American immigration. They would be the celebrities of the time, as their personal lives made headlines. Their huge furnaces and factories were a matter of national pride and support of American exceptionalism. They would give rise to great senators and congressmen such as Pennsylvania's "Pig Iron" Kelly, and in 1896, they would elect one of their own as President of the United States, William McKinley. And years later a group of aging Pig Iron Aristocratsmet in a room in Youngstown and selected William Taft to be the Republican candidate for the presidency in 1908.

CHAPTER TWELVE

Battle of the Furnaces

The early 1870s would bring ironmaking and blast furnaces to the front pages of the iron district newspapers. The great blast furnaces of Youngstown and the Mahoning Valley dominated the 1850s and 1860s. In 1846, the famous Eagle Furnace of Youngstown set a record of twenty-six tons of pig iron in a day. The Eagle Furnace used heating stoves to heat the air blast, which is typical of today's great blast furnaces. That Eagle Furnace record stood for nine years, until Youngstown's Falcon Furnace set the new record at fifty tons in a day. That record fell in 1868 to Youngstown's Haselton No. 2 with sixty tons in one day. In 1869, John Struthers built the world's largest blast furnace, known as Anna Furnace, at Youngstown. It stood seventy-five feet high and had a sixteen-inch bosh. Using Brier Hill coal it set the monthly tonnage record of over 1,600 tons in a month. It also used Lake Superior iron ore supplied by Pickands and Mather. The achievement took both the Youngstown and Pittsburgh headlines. Pittsburgh's forty-five-feet Eliza and Clinton Furnaces were dwarfed by comparison. These achievements and the availability of lake ores stirred the dreams of Pittsburgh's Pig Iron Aristocrats.

While the Youngstown district was far ahead of Pittsburgh after the war in pig iron manufacture, Pittsburgh remained the center of pig iron use in rolling mills and foundries. Youngstown had the advantage of lake ore and coal. There was also a new breed of Pig Iron Aristocrat in Pittsburgh. These new Pig Iron Aristocrats were moneymen like the Pig Iron "Jupiter," James O'Hara, of earlier times. Henry Oliver (1840-1904) was one of this new breed. He had immigrated to Pittsburgh from North Ireland in 1842. Oliver would eventually fit the old mold well, being a Presbyterian, Scots-Irish, a soldier, and a Mason. He would, however, start life in the Scottish ghetto on Pittsburgh's north side, six years earlier than Carnegie but a few years younger in age. In 1849, Henry Oliver entered the employ of the Atlantic and Ohio Telegraph Company. He would work with several other future greats. One was a young Andrew Carnegie; another was Robert Pitcairn, later superintendent of the Pennsylvania Railroad. As a telegraph delivery boy, Oliver learned the business of Pittsburgh. In 1859, he became a clerk at the Clinton Furnace of the firm of Graff, Bennett, and Company.

Oliver moved rapidly from a clerk to an iron product salesman. He saved money as a religion and invested, as did his friend Andrew Carnegie, in Pittsburgh business. He made a large profit in steamboat companies, and in 1865 invested in the city's prime business: pig iron. Oliver had become familiar with the firm of Lewis and Phillips, since this bolt-making firm purchased iron bar from Clinton Furnace. William Lewis was the inventor of the first bolt-making machine. After a couple years as a volunteer soldier, Oliver invested in Lewis's firm, forming the company of Lewis, Oliver, and Philips. The firm boomed during the war and expanded with three rolling mills in the Pittsburgh area. Maybe just as important was the young Oliver's work in the new Republican Party of Pittsburgh. In the Lincoln election of 1860 Pittsburgh and Pennsylvania broke the chains of the 1794 Whiskey Rebellion and turned against the Democrats. The victory was really one for the tariff platform of the Republican Party. The Pig Iron Aristocrats finally persuaded the population of the importance of protected industries for economic health. In particular, it inspired a lifelong commitment of these young Pig Iron Aristocrats to the Republican Party. The Duquesne Club would combine Scots-Irish Presbyterianism, Monongahela Rye, and business. It would grow into a castle of pig iron capitalism. At the Duquesne Club, the Pig Iron Aristocrats met daily in "Room Number 6." These old-line industrialists included Benjamin Jones, Henry Oliver, Henry Phipps, C. B. Herron, J. W. Chalfant, and C. H. Spang. Oliver had helped found the aristocratic club of the pig iron manufacturers, the Duquesne Club. Such meetings today would border on illegal.

Oliver was a member of two Scots-Irish gangs of Pittsburgh's north side that would change the future of pig iron and usher in the Age of Steel. One gang in the late 1850s was the "original six," which consisted of Andrew Carnegie, Thomas Miller, William Cowley, James Smith, James Wilson, and John Phipps (brother of Henry Phipps). The younger gang consisted of Tom Carnegie, Henry Oliver, Henry Phipps, and Robert Pitcairn. In the 1870s, these names would change the pig iron industry, create the steel industry, and change American business forever. The story of the Carnegie Empire and its impact on the pig iron industry would begin with two Prussian brothers and blacksmiths in 1858. Andrew's group would be the core of a new breed of Pittsburgh ironmakers.

Pittsburgh's need for pig iron had had almost tripled during the Civil War; furthermore, war inflation had taken the price from eighteen dollars per ton to forty dollars per ton. Most of the pig iron was coming from Youngstown's new blast furnaces and Ohio's Hocking Valley charcoal furnaces. Pittsburgh needed local and economic pig iron supplies to maintain its massive foundries and rolling mills. Two Pittsburgh furnace projects were under consideration by various Pig Iron Aristocrats. One of these was a partnership of Lewis Dalzell and Company, Painter and Sons, Graff Bennett and Company, Spang, Chalfant, and Company, Oliver Brothers, and Smith Foundry. This pig iron combination was a result of the meetings in Room Number 6 of the Duquesne Club. Union Iron Mills was asked to join as well. Union Iron, however, being the largest user of pig iron, decided it would need a furnace dedicated to supply them. Even without Union Iron, the partnership known as the "aristocracy"[15] moved ahead, forming Isabella Furnace in 1872. Isabella was seventy-five feet high with an

---

[15] Peter Krass, *Carnegie* (Hoboken: John Wiley & Sons, 2002), 169.

eighteen-inch bosh. The original goal was five hundred tons of pig iron a week.

At the same time a Carnegie group moved with a furnace to supply the Union Iron Mills. They organized the firm of Kloman, Carnegie and Company to build Lucy Furnace (named after Tom Carnegie's wife). Andrew and Anton (changed to Anthony) Kloman were Prussian blacksmiths who had opened a forge shop a few miles up the Allegheny River from Pittsburgh in 1858. The forge was capable of making large wagon axles and heavy railroad axles. The growth was rapid for such a large forge. By the middle of 1859, they needed money to invest in a second trip hammer. This brought the brothers in contact with Tom Miller and Henry Phipps of the "original six." Miller and Phipps not only invested in the second trip hammer, but also secured a major line of business producing axles for the Pittsburg, Fort Wayne, and Chicago Railroad (then called the Ohio and Pennsylvania). The Civil War created a huge demand for heavy axles, increasing the price from two cents a pound to twelve cents a pound. In 1862, the firm built a massive plant on the north side of Pittsburgh, and became a major user of pig iron. By 1860, the plant was known as Kloman-Phipps Iron City Forge. The Carnegie brothers entered the picture when the they allied with another plant of the Kloman-Phipps known as Cyclops. The merger of Cyclops with Carnegie's Union Iron created what was known as the Lower and Upper Union Mills. The Kloman partnership led to further joint ventures, agreements, and mergers with Carnegie. These combinations also created a new breed of Pig Iron Aristocrats in the Kloman brothers, Carnegie brothers, Henry Phipps, and Thomas Miller. The new group was in direct competition with the old-line aristocrats of Oliver, Dalzell, Spang, Chalfant, Jones, and Laughlin. These combinations created the biggest demand for pig iron in the region, requiring a separate local pig iron furnace to feed them.

The duel of Isabella and Lucy Furnaces started in June of 1872. The battle became the headlines of Pittsburgh's papers as the furnaces overtook the Struthers Furnace in Ohio. For the first year, the weekly tonnage record moved back and forth between the two furnaces. Lucy Furnace ended the year with a five hundred-ton a week average versus a 498-ton a week average for Isabella. The battle continued for years, as the international press started to follow it. In 1874, Lucy set the daily world production record at one hundred tons a day.

Demand for pig iron soared as Andrew Carnegie brought on the Bessemer steel mill of his Edgar Thomson Works in 1876. Edgar Thomas Works had to depend on pig iron supply from Lucy Furnace, lacking any blast furnaces of its own. Pig iron was loaded daily on the Pennsylvania Railroad and shipped a few miles up the Monongahela River to Edgar Thomson Works. There the pig iron was re-melted to feed the Bessemer steel converters of Edgar Thomson Works. In 1876, using one furnace, Lucy Furnace produced 16,174 tons of pig iron. In 1878, using two furnaces, Lucy Furnace produced 62,102 tons. In 1881, Isabella set the daily record at one thousand tons per day. Both plants rebuilt and added new furnaces during the fifteen-year battle. The battle even got very personal between the men and managers. The battle and fierce competition created new furnace technology, improved mining in the Michigan ore mines, and improved coke manufacture. In 1880, American pig iron production overtook that of England to lead the world. This accomplishment was driven first by the use of coke, which overtook charcoal fuel in 1870 and anthracite coal in 1880. The other factor was the rich lake ores, which could arrive by canal, river, or railroad. England was on its way to becoming a "nation of merchants," while America would soon become the world's manufacturing empire.

The Pittsburgh furnace battle of Lucy and Isabella Furnaces would become one of chemistry and metallurgy. Both of these sciences were brought to new levels of importance. In particular, Carnegie partner, Henry Phipps, was interested in industrial chemistry and metallurgical engineering. Phipps lived nearby and was known to visit the furnace day and night. Lucy Furnace was a technical marvel. Its height was achieved by using a cast iron shell that was lined with firebrick. Furnace engineer Mr. Whitwell had applied improved heating stoves of his own design. Whitwell invented a "bell" to better distribute the load of ore. He also built a glass model of the furnace to study furnace loading and better practices. Whitwell and Phipps also implemented the use of puddler furnace cinder, which had an iron content near that of iron ore. Puddler cinder was a waste product at the Carnegie group's Union Mills. Using puddler cinder for twenty-five percent of the furnace burden allowed Lucy Furnace to maintain a commercial advantage over the competition. The advantage showed that huge savings were possible by owning integrated operations. Phipps would later encourage a young Charles Schwab in chemistry; Schwab would become the first president of United States Steel in 1901.

Lucy Furnaces grew in productivity and improvements because of their key supply role to the Carnegie group. In 1875, Carnegie's huge new Bessemer steel plant came on line in Braddock, Pennsylvania, known as Edgar Thomson Works. Edgar Thomson Works became the major customer of Lucy Furnaces, but it was an inefficient arrangement. Pig iron was cast at Lucy and shipped by rail to Braddock. It was then remelted in a cupola furnace to feed the Bessemer converter. In 1879, Mahoning Valley furnace wizard Julian Kennedy was brought to Braddock to build a furnace. The basic structure was an old charcoal furnace transferred from northern Michigan. The Kloman brothers had owned the old Escanaba furnace. In 1880 this rebuilt furnace set a world record for low coke consumption. Within a year there were two furnaces setting world production records.

But in 1882, the plant manager of Carnegie's Edgar Thomson Works, Captain Bill Jones, would change the pig iron industry and usher in the integrated steel industry. Until Bill Jones invented the "Jones Mixer," the pig iron industry was a stand-alone one. A blast furnace produced and cast pig iron units. These fifty-pound pigs were moved cold to another furnace to re-melt the pig iron for foundries or Bessemer converters in steel plants. The Jones Mixer allowed hot liquid metal from several blast furnaces to be held and transported as a liquid to a Bessemer converter. It resulted in savings and allowed for a continuous operation. Integrated steel companies now started to build their own furnaces as part of the steel plants. Single pig iron furnace operations became dependent on foundries. The Pig Iron Aristocrats slowly converted to steel masters, but the political name remained. Carnegie was forcing the conversion not because of process integration, but because steel was overtaking the use of puddled iron in many applications, such as railroad rails. Jones and Laughlin in Pittsburgh, and Cambria Iron in Johnstown, Pennsylvania, produced most of the domestic iron rails by 1870, and the Republicans had blocked the earlier British monopoly on iron rails. But in 1875, Andrew Carnegie was betting on steel to replace wrought iron and that threatened the livelihood of the old iron aristocrats. Still, stand-alone pig iron production would remain ahead of integrated steel production until after 1900.

CHAPTER THIRTEEN

The Peak and Decline of the Pig Iron Aristocracy

The Pig Iron Aristocracy would peak in the 1890s, when integrated steel manufacture would overtake pig iron production. The legacy of the Pig Iron Aristocrats was more political than technical. The Pig Iron Aristocrats had given rise to the Republican Party and the application of protectionism. The Pig Iron Aristocrats managed the United States ascendancy as the world's great manufacturer. They had battled down the free trade Democrats, and had passed the McKinley Tariff of 1890. The Democrats' main argument was that protection would result in price fixing by the domestic producers. Andrew Carnegie would summarize the theory of protectionism in 1882: "We are creatures of the tariff, and if ever the steel manufacturers here attempt to control or have any general understanding among them the tariff would not exist one session of Congress. The theory of protection is that home competition will soon reduce the price of this product so it will yield only the usual profit. Any understanding among us would simply attempt to defeat this. This never has been or ever will be such an understanding." These Pig Iron Aristocrats realized that the massive political support and their labor alliance depended on their re-investment into American plants. The whole alliance depended on a commitment to American nationalism. The merger of the Iron Whigs of Ohio, the New England industrialists, the

Pittsburgh capitalists, old Henry Clay supporters, and the northern abolitionists created the Republican Party.

Protection of America's pig iron industry had given the Pig Iron Aristocrats a new political party. They assured the success of its young candidate Abe Lincoln, and by doing so launched a war, created a new state, and took control of American politics. In Ohio and Pennsylvania, new political rings and city machines arose. They were modeled after the earlier Democratic Party machines such as Tammany Hall. In Pittsburgh, men like Christopher Magee and William Flinn became political bosses. These bosses played a critical role for the Pig Iron Aristocrats. The Aristocrats would buy some politicians, but could not assure the vote delivery needed. The bosses acted as brokers in the process. Votes in the pig iron districts could be counted on and presidential candidates had to cater to these districts. Protectionism united the aristocrats, workers, and bosses. It was a perfect symbiotic relationship in many cases. These county machines of the Aristocracy in Pittsburgh's Allegheny County, Ohio's tri-county area of the Mahoning Valley, Ohio's Hocking Valley counties, eastern Pennsylvania's coal counties, and New England's Connecticut Valley could deliver a big enough majority to affect presidential elections. These areas saved the second election of Abraham Lincoln, which preserved the Union.

The Pig Iron Aristocrats foreshadowed the robber barons in their philanthropy. They followed the tradition set in the 1790s by James O'Hara. They built parks, donated churches, and contributed to cultural endeavors. William Thaw, Henry Buhl, and Henry Clay Frick contributed to the building of Allegheny Observatory in the 1860s. William Thaw, even before his son, had supplied the first funds in the 1860s. Henry Buhl donated a planetarium and Henry Phipps a conservatory. Christopher Magee built the Pittsburgh Zoo. Thomas Mellon gave to the University of Pittsburgh for an Institute of Industrial Research. Their donations built a city of churches. All of this was before Andrew Carnegie came on the scene. William Thaw is of particular note because he was Pittsburgh's king of transportation, which allowed the iron industry to prosper. In the 1840s, Thaw had teamed up with Benjamin Jones and Samuel Kier to bring the canal system to Pittsburgh. The canal link would ultimately tie the great iron ore deposits of Michigan to the area. Politicians were also made and backed by these Pig Iron Aristocrats. The boss system, while inefficient and corrupt, did offer social services to the poor. Many of the manufacturing cities of the east such as Boston and Philadelphia also had Republican machines.

Men like William Thaw, Harry Thaw (son of William), Benjamin Jones, James Laughlin, John Shoenberger, E. Stevens, Henry Phipps, Henry Oliver, and Henry Buhl were the Pig Iron Aristocrats of Pittsburgh. Actually, many lived on Pittsburgh's north side, then known as Allegheny City. In the 1850s, the original "Millionaire's Row" was Allegheny City's Ridge Avenue. No city in America had such wealth concentrated in one row of mansions. Allegheny City soon became the classy suburb of Pittsburgh industrialists. Lincoln would even make a brief stop there prior to going to his hotel in Pittsburgh on his way to the White House. Presidents Ulysses S. Grant and Rutherford Hayes would also visit this aristocratic city of pig iron manufacturers; actually every president from Lincoln to Taft would visit Pittsburgh.

At the heart of pig iron politics was protectionism and nationalism. The iron districts had supported Henry Clay and Republicans through President Harding. The election of Lincoln was the first big national triumph of the Pig Iron Aristocrats. The Lincoln victory was based on his belief in protectionism, which brought the Iron Whigs into the Republican Party. The alliance of the Iron Whigs, Quaker abolitionists in the Western Reserve of Ohio, and Pig Iron Aristocrats of Pennsylvania and Connecticut was the great alliance that also ended slavery. Lincoln would inspire a long line of Republican protectionists such as "Pig Iron" Kelly and William McKinley. President McKinley did much more than implement the ideas of Henry Clay and Alexander Hamilton; he added to and improved the "American System." "Pig Iron" Kelly certainly influenced McKinley's additions to the American System. "Pig Iron" Kelly represented the Philadelphia district for almost twenty years; McKinley was his protégé on the House Ways and Means Committee. Kelly trained him and eventually assured his position as Chairman of the Ways and Means Committee after Kelly stepped down. Kelly influenced McKinley in seeing that Democracy and Capitalism are interrelated in America; they combine to form the "American System." Kelly had championed black suffrage because equality and opportunity are the foundation of American capitalism. Capitalism without opportunity, upward mobility, and civil rights is no better than tyranny. Kelly had also fought for the shorter workweek and better working conditions, as would McKinley. McKinley incorporated all of this into his vision. It was Clay's system adapted for the Industrial Revolution.

On a national level, the Pig Iron Aristocrats delivered a long line of protectionist presidents including iron district ones like Garfield and McKinley. The greatest of the kingmaker Pig Iron Aristocrats was Marcus Alonzo Hanna of Cleveland. Hanna was a high school classmate of John D. Rockefeller, started in oil refining in the late 1860s. Over the next two decades he moved into lake transportation, coal mining, iron mining, and banking. He emulated the career of James O'Hara in early Pittsburgh. He grew rich with the iron ore trade to Cleveland, and his business empire stretched from northern Michigan to the blast furnaces of Youngstown and Pittsburgh. In the 1880s, he moved into politics, and became a leader in Ohio's Republican Party. Hanna used his pig iron network to raise money for Republicans. He purchased a newspaper to help build his political power. He was the campaign manager for Ohio's Senator Sherman. He then helped William McKinley win as governor of Ohio in 1891. In 1894, he became the campaign manager for William McKinley, and forged the labor-capital alliance. Hanna also formally tied the iron districts together into a political block. He worked the great iron triangle of Ohio and Pennsylvania for votes and money. He was the first to fully realize the power of the Pig Iron Aristocracy as a true national force. He awoke the old iron protectionist network of Henry Clay.

The last great boom for the Pig Iron Aristocrats would be the benefit of the McKinley Tariff of 1890. The American tinplate

The Peak and Decline of the Pig Iron Aristocracy   153

industry became the poster child for this new approach of tariff legislation. Tinplate was rolled iron covered with tin. America's demand for tinplate was enormous, with the tin can market for food in 1889 being twenty-nine million dollars alone, and with the oil industry providing an even greater demand for large tin cans (refined kerosene being sold in tin cans). In 1889, there wasn't a single tinplate mill in the country. England, because of its large tin deposit at Cornwall, had a world monopoly on tinplate. McKinley, working with ironmakers, believed that with a tariff of around fifty percent, a domestic industry could be developed. Manufacturers from Wheeling, West Virginia; Youngstown, Ohio; and Pittsburgh, Pennsylvania, were highly supportive. Old friends and rolling mill owners in Niles lobbied McKinley on the industrial potential. The Democrats fiercely objected to this type of industry building by tariff reform, arguing that it would mean years of high-price tinplate before any results could be seen. The political battle would be extremely bitter. The passage of the Tariff of 1890 would cost McKinley his seat in Congress, but would energize the Pig Iron Aristocrats nationwide to give him the Presidency. America would also have to import or develop a new technology. In addition, new mines of American tin would have to be developed. Finally, Democrats argued it favored a small group of manufacturers at the cost of all Americans. This approach to use tariffs to promote specific industries was novel at the time, but had roots back to Henry Clay. The bill did provide an escape clause if American manufacturers failed to make the necessary investments. If American manufacturers could not maintain a third of the market demand, then tinplate would become duty-free again. If success were not obtained by 1896, the tinplate tariff would be eliminated. Manufacturers, however, did respond, providing millions for investment. They imported thousands of skilled tinplate workers to develop the technology as well. By 1895, there were sixty-nine American tin mills, and they would be a major consumer of pig iron in America.

The Democrats in 1890, however, seized the moment, by supporting McKinley for high tariffs on tinplate, which was the base for the farmer's tinware. The idea was to drive prices up and cause a political backlash for McKinley and the Republicans. McKinley's plan would take time for the iron industry to invest, and that would give the Democrats a window of opportunity. In tinplate, America was dependent on outside sources such as Britain, as Americans, scrambled to make the investments. Tinplate was an embryonic industry that McKinley felt critical for the American economy, but it would need at least two years before it could stand on its own. The tariff had focused on developing domestic tinplate to help reduce prices of tinware down the road. In any case the merchants and peddlers did increase prices in anticipation of a price increase. In addition, Ohio peddlers, who resented the tariff, started to increase prices aimed at affecting the 1890 election. The actual tariff increases of the 1890 bill would take place after the election. The Democratic candidate even paid peddlers to raise prices. Ohio peddlers even got encouragement (and probably help) from the Democrats to set prices as high as one dollar a pot and twenty-five cents a cup; prices which would normally be twenty-five cents a pot and five cents for a cup. The prices were beyond what the farmers could pay, and though the peddlers didn't sell any, they achieved their goal. As farmers complained, they were told to blame it on the McKinley tariff. The issue was completely bogus with tinplate, and a Grange member noted, "I do not know of a single article that is higher than a year ago. Of course, it was less politic to say so from the stump."[16] Historical statistics support that there was no evidence of overall price increases.[17]

McKinley lost his seat in Congress in his three-county district, which was just outside the pig iron strongholds a few miles east. Across the nation in iron districts, there was disbelief that McKinley lost after his greatest achievement. McKinley had stood true to his ironmaker roots, and the Pig Iron Aristocracy would not forget.

---

[16] J. H. Brigham to A. J. Rose, September 4, 1891, Rose papers, Texas History Collection.
[17] Albert Rees, *Real Wages in Manufacturing 1890-1914* (Princeton: Princeton University Press, 1961), 74-77.

History would also vindicate McKinley and the Pig Iron Aristocracy, as a forty-million dollar American industry was in place by 1896, generating tens of thousands of American jobs. The response of American steel companies represented a manufacturing miracle, but too late to help in the 1890 election. By the end of 1891, the tinplate industry had twenty plants operating and ten more under construction. By the deadline of 1896, there were seventy-five plants in operation. Furthermore, the great industrial towns of Youngstown and Wheeling grew as a result of the tariff. Many credit the logistical successes of American "doughboys" in World War One to the existence of large tinplate industry due to the McKinley Tariff.

Cleveland pig ironmaster and politician Mark Hanna would lead the charge for William McKinley's presidential campaign. Hanna had inspired a beaten congressman to run for governor of Ohio, which McKinley won and held until his presidential run in 1896. McKinley as Ohio's governor was a popular speaker in the iron and steel districts of Illinois and Pennsylvania. Hanna proved a true visionary in campaigning. Historian Lawrence Goodwin noted: "In sheer depth, the advertising campaign organized by Mark Hanna in behalf of William McKinley was without parallel in American history. It set the creative standard for the twentieth century."[18] He was one of the first to use political polls, and their use helped him identify an early boom for Bryan, the Democratic candidate. He was the first to use the telephone extensively, paying for lines where needed. Hanna used news summaries to stay informed on editorial comment across the nation. Never before had literature been used so successfully. Earlier elections had focused on political speakers, but Hanna found pamphlets to be almost as effective. Hanna printed the pamphlets in many languages. Hanna used the mill owners and managers to get the pamphlets in the hands of workers. He brought in speakers who spoke the native tongue of the workers, and advertised in specially nationality newspapers. The tariff issue

---

[18] Lawrence Goodwin, *The Populist Moment* (Oxford: Oxford University Press, 1978), 282.

played well with the industrial workers, and pamphlets added focus and local color.

Hanna pooled the money of men such as Henry Clay Frick and the Ohio ironmasters. Polls also helped Hanna catch the shift towards McKinley in late August. The shift allowed McKinley to hammer home the tariff issue, to which the Democrats did not respond. McKinley and Hanna used regional speakers in a masterly way. These speakers were often senators and representatives, a role McKinley had played for Hayes, Garfield, Harrison, and Blaine. McKinley used Terence Powderly, former Grand Master of the Knights of Labor, to maintain his edge with labor. Bryan lacked the money to put speakers in the field, and depended too much on himself to carry the message.

The death of William McKinley brought a true end to the Pig Iron Aristocracy, which passed the torch to the Steel Trust. The first sign of the end of the Pig Iron Aristocracy came in 1901 with the formation of United States Steel Corporation. Andrew Carnegie signed over his steel company to New York banker J. P. Morgan. This merger would cobble together some of the great mills of Pittsburgh and Chicago. The control of iron and steel production passed from the operators to the financial centers of New York. The Pig Iron Aristocrats had always been operators, not bankers. They were men that handled the metal. The new Steel Age would be different. Bankers would form the new companies, and many would not even be able to identify pig iron. These businessmen were international in their view, with less interest in protecting American workers and plants. The Pig Iron Aristocrats were nationalists and Americans. While far from perfect, they had a closer relationship with American labor because of their domestic strategy. It wasn't all about the money; it was the power and might of the industry.

September 5, 1901, was to mark the final phase of a vision that had built an American industrial empire. The bulldog of protectionism had just completed a notable speech at the Buffalo Pan-American Exposition, which most agree was the best of his long career. The next day at the Exposition, an anarchist and former steel worker felled McKinley with a bullet that would eventually kill him. It heralded a significant change in America's view of industrial America's role in the world. The McKinley trade tariffs had brought an unheard-of level of prosperity to the Pig Iron Aristocrats and American industry in the last quarter of the nineteenth century. The problem now was the American market in 1901 alone could not drive the industrial growth of the past thirty years. That growth was the result of the Pig Iron Aristocrats' determination to protect, expand, and build a lasting legacy for American industry. From his youth, William McKinley had championed the belief of Henry Clay in industrial republicanism, which blended Jeffersonian creativity, Emerson's self-reliance, and Hamiltonian finance, with Washingtonian patriotism. American industry knew no rival and accepted no competitor under the McKinley Tariff. McKinley, the Republicans, and the Pig Iron Aristocrats' protectionism was unabashed Americanism. McKinley viewed it more as an economic Monroe Doctrine than protectionism. It was a political alliance of big business and the American workingman to secure America's industrial destiny. He believed the next world empire would not be one of militaristic strength, but one of industrial might. It reflected a global, economic manifest destiny. To the cheers of the American public that day in Buffalo, McKinley prepared to launch the final phase of industrial conquest: world markets.

September 6, 1901, ended in many ways the heroic view of American industrialism. McKinley struggled for days at a private home in Buffalo, but the end came on September 14. September 14, 1901, was also the thirty-ninth anniversary of the Battle of South Mountain where he had distinguished himself for bravery in the Civil War. McKinley's last words were representative of his life- "Good-by, all; good-by. It is God's way. His will be done." He died a man of great faith and deep love for his country. He saw the best in people to a fault, often being let down. While a socialist bullet had ended his life, McKinley believed that America was a nation destined to lead the world with its capitalism. Unlike those who would follow, he never feared socialism as a fundamental alternative to the American System, but as a plague of those who put greed above nation. He even saw his assassin as misguided versus being an enemy.

With the death of McKinley and the formation of United States Steel Corporation, the last stronghold of the Pig Iron Aristocracy would be the Youngstown-Niles area. It would be here that they maintained their tinplate mills. Pig iron remained king in the Mahoning Valley until 1894, when the first steel plant opened. It was in Youngstown that a small group of aging Pig Iron Aristocrats helped pick the Republican nominee and future president – William Taft. The next year, President Taft dedicated the Niles McKinley Memorial. The bronze busts included many of the Pig Iron Aristocrats including Jonathan Warner, pioneer in bringing lake ore to the Mahoning Valley and furnace pioneer Alexander Crawford.

CHAPTER FOURTEEN

The Legacy

The Pig Iron Aristocracy really represented some of America's best capitalists. They had worked their way up from humble beginnings. They loved the production of pig iron as much as the profits. They used politicians and bankers to that end. When their politicians passed protective tariffs, they poured their extra profits back into plant improvement and expansion. They felt uncomfortable with the rise of huge corporations, preferring the mythical past of the iron plantation master. Money changed them little. They lived in the smoky cities of their factories. They gave to their communities without the fanfare of the corporate steel barons that followed them. They had an inherent love of capitalist competition and rarely tried to control price or form trusts. They promoted men based on their ability to produce more pig iron or make it more efficiently. While they were ruthless competitors at times, they were Americans first. They, at times, were far from generous with wages, but they were committed to increasing American jobs. While they fiercely opposed unions, they often practiced a paternal approach to their workers going back to the model of the iron plantation masters. They gave us one of the most scandal-free presidents in our history in William McKinley.

Today they are almost completely forgotten except for a few street names and aging buildings. It is doubtful at even one American school child today could recognize the names of Jones, Laughlin, O'Hara, Craig, Struthers, Whitaker, Shoenberger,

Meason, Phipps, Oliver, Buhl, Thaw, Mather, Tod, and Stevens. Their bronze busts can still be seen in the aging McKinley Memorial at Niles, Ohio, yet their legacies can be found in the traditional strength of the Ohio Republican Party. Pig iron production remains, but its peak as an engineering material has long ago passed to the steel industry. And the American steel has passed its torch to other nations. The last bastion of the Pig Iron Aristocracy was the Niles-Warren-Youngstown area of Ohio.

The Greek architecture and the bronze statues of the Niles Memorial related to my own vision of the period. It was striking to see a bronze statue of fellow Pittsburgher, Henry Clay Frick, in such a prominent position as the major donor. In Pittsburgh, Frick still represents the evil of capitalism and still brings anger in the hearts of many union leaders. Other Pittsburghers such as George Westinghouse, Andrew Mellon, Benjamin Jones, Henry Oliver, A. M. Byers, Philander Knox, and the "Boys of Braddock" (Carnegie Veterans Association) were there. It was in my research on biographies of these Pittsburghers that McKinley's name was so prevalent. There were the busts of many Mahoning Valley steelmakers and Ohio industrialists such as Mark Hanna. As a metallurgist, I recognized some greats in bronze such as John Battelle, who rolled the country's first tinplate and founded Battelle Institute in Columbus where I had once studied. The busts include Mahoning Valley ironmakers Thomas Struthers, David Tod, Joseph Brown, Richard Brown, James Ward, Joe Butler, Andrew Crawford, Jonathan Warner, and James Heaton. There are also the busts of the steel men who honored these old aristocrats such as Andrew Carnegie, John "Bet-a-Million" Gates, James Farrell, Judge Gary, and George Baker. The presidents include William Howard Taft and Theodore Roosevelt. Supreme Court Justice William Day is also represented. It is this Pantheon of Pig Iron Aristocrats that still stands in testimony to their greatness.

Bibliography

Baldwin, Leland. *Pittsburgh: The Story of a City*. Pittsburgh, Pennsylvania: University of Pittsburgh Press, 1937.

Bell, Thomas. *Out of this Furnace*. Pittsburgh, Pennsylvania: University of Pittsburgh Press, 1976.

Boucher, John. *William Kelly, A True History of the So-Called Bessemer Process*. Greensburg, Pennsylvania: By Author, 1924.

Bridge, James. *The Inside History of the Carnegie Steel Company*. New York, New York: Aldine Book Company, 1903.

Buck, Solon. *The Planting of Civilization in Western Pennsylvania*. Pittsburgh, Pennsylvania: University of Pittsburgh Press, 1968.

Carnegie, Andrew. *The Autobiography of Andrew Carnegie*. Boston, Massachusetts: Northeastern University Press, 1920.

Cason, Herbert. *The Romance of Steel*. New York, New York: A.S. Barnes & Co., 1907.

Cotter, Arundel. *The Authentic History of the United States Steel Company*. New York, New York: Moody Book Company, 1916.

Davison, Mary. *Annals of Old Wilkinsburg and Vicinity*. Wilkinsburg, Pennsylvania: The Group, 1940.

Eggert, Gerald. *Steelmasters and Labor Reform, 1886–1923*. Pittsburgh, Pennsylvania: University of Pittsburgh, 1981.

Fisher, Douglas. A. *The Epic of Steel*. New York, New York: Harper and Row, 1963.

Fisher, Douglas. *Steel Serves the Nation*. Pittsburgh, Pennsylvania: USS Corporation, 1951.

Frick Symington Sanger, Martha. *Henry Clay Frick*. New York, New York: Abbeville Press, 1988.

Harvey, George. *Henry Clay Frick*. Pittsburgh, Pennsylvania: Charles Scribner's Sons, 1928.

Hessen, Robert. *Steel Titan: The Life of Charles M. Schwab*. New York, New York: Oxford Press, 1975.

Holbrook, Stewart. *Iron Brew*. New York, New York: The Macmillan Company, 1939.

Holbrook, Stewart. *Age of the Moguls*. New York, New York: Doubleday, 1954.

Kopperman, Paul. *Braddock at the Monongahela*. Pittsburgh, Pennsylvania: University of Pittsburgh Press, 1977.

Krass, Peter. *Carnegie*. Hoboken, New Jersey: John Wiley, 2003.

Lamb, George. *The Unwritten History of Braddock's Field*. Pittsburgh, Pennsylvania: Nicholson Printing, 1917.

Livesay, Harold. *Andrew Carnegie*. Boston, Massachusetts: Little, Brown and Company, 1975.

Lorant, Stefan. *Pittsburgh: The Story of an American City*. Lenox, Massachusetts: Authors Limited, 1964.

Newton, John. *A Century and a Half of Pittsburgh and her People*. Pittsburgh, Pennsylvania: Lewis Publishing, 1908.

McCullough, David. *The Johnstown Flood*. New York, New York: Touchstone Books, 1986.

Rupp, Daniel. *Early History of Pennsylvania and the West*. Pittsburgh, Pennsylvania: D. W. Kaufman, 1846.

Schom, Alan. *Napoleon Bonaparte*. New York, New York: HarperCollins, 1997.

Serrin, William. *Homestead*. New York, New York: Times Books, 2003.

Sharp, Myron and William Thomas. *A Guide to the Old Stone Blast Furnaces in Western Pennsylvania*. Pittsburgh, Pennsylvania: The Historical Society of Western Pennsylvania, 1966.

Stubbles, John. *The Original Steel Makers*. Warrendale, Pennsylvania: Iron and Steel Society, 1984.

Swetnam, George. *Andrew Carnegie*. Boston, Massachusetts: Twayne Publishers, 1980.

Temin, Peter. *Iron and Steel in Nineteenth Century America.* Cambridge, Massachusetts: M.I.T. Press, 1964.

Vogt, Helen. *Westward of ye Laurall Hills.* Parsons, West Virginia: McClain Printing, 1976.

Walkinshaw, Lewis. *Annals of Southwestern Pennsylvania.* New York, New York: Lewis Historical Publishing, 1939.

Winkler, J.A. *Incredible Carnegie.* New York, New York: The Vanguard Press, 1931.

Index

23rd Ohio Regiment, 128
Abolitionists, 151
Adams, John Q., 62, 65, 80
African iron reduction process, 5
Age of Steel, 7
Agnew, David, 101, 114–115
Agrarianism versus Industrialization, 55, 58
Albert, Prince, 69
Algers Foundry, 133
Alleghany Furnace, 107
Allegheny Observatory, 53
Allen, Ethan, 12, 16, 100
Alliance Furnace, 16, 45
Allison, Nancy, 90
American Industry League, 70
American Iron Works, 124–125
American Pig Iron Association, 91
American Steel and Wire, 52
"American System", viii, ix, 19, 55, 57, 60, 62–63, 65–66, 68–70, 80, 84, 98, 115, 127, 151–152, 157
Ames family, 122
Ames Foundry, 133
Ames Ironworks, 132
Anderson, William, 41
Anna Furnace, 141
Anshultz, George, 41
Anshultz Furnace, 38
Anthracite coal, 74, 121
  in eastern Pennsylvania, 107
  in Scotland, 107
  properties, 107
Anti-Masonic Party, 80
Anti-Slavery, 137
Armor plate production, 136–137
"Arsenal of the Revolution", 12, 100
Arthur, Chester, 19

Arundel iron ore formation, 10–11, 113
Atlantic and Ohio Telegraph Company, 142
Audubon, John, 24
Bailey, William, 115–116
Bailey, Woodward and Company, 116
Baird, Thomas, 48
Baker, George, 160
Bakewell
  Benjamin, 77
  Thomas, 80
Bakewell family, 81
Baldwin, Henry, viii, 45, 57, 61, 77
Baltimore and Ohio Railroad, 110
Bank of Pennsylvania, 46, 48, 74, 78
Bank of Philadelphia, 107
Bank of Pittsburgh, 81–82
Bank of the United States, 81–82
Barker, Joseph, 48
Battelle, John, 160
Bay Furnace, 101
Bayard
  Samuel, 37
  Stephan, 36
Bayard family, 46
Beelen, Anthony, 48
"Beelzebub" twenty-inch cannon, 136
Belknap, M. B., 50
Belmont Iron Works, 116, 137
Benton, Thomas, 80
Bessemer
  conversion process, 8, 146
  Henry, 7
Bessemer steel, 145
Bird family, 15

Bituminous coal, 108, 126
  converted to coke, 107
  properties, 107
  from western Pennsylvania, 107
"Black-band" ore, 92
Blacksmiths, guilds and spiritual rituals, 2
Blaine, James, 80, 82, 156
Blanchard, Thomas, 99
Blast furnaces, 120
  evolutionary rise, 121
  Union, 128
  used today to produce pig iron, 5
Bog iron, 11
Bog ore, 4, 89, 100
Boone, Daniel, 19, 27, 29
Boys of Braddock, vii, 160
Brackenridge
  Henry, 23
  Hugh Henry, 37, 75
Brackenridge family, 46
Braddock, Edward, 29–30, 39
Braddock Road, 29
Braddock's Defeat, 29
Breecher, Henry Ward, 80
Brier Hill coal, 92–93, 141
Brier Hill farm, 92
Brier Hill Iron and Coal Company, 93
Bronze Age, 4
Brown
  Joseph, 95, 123, 160
  Richard, 160
Brownstown, 124
Bryan, William Jennings, 155
Buchanan, James, 19
Buckland, Cyrus, 99
Buena Vista Furnace, 132
  first blast furnace, 122
Buffington's Island, battle of, 128
Buhl, Henry, 138, 150–151, 160
Burd, James, 29

Butler
  Joe, 160
  Joseph Jr., viii, 90–91, 95
  Joseph Sr., 90–91
  Richard, 36
  William, 36
Byers, A. M., 160
Calhoun, John, 19, 59
Cambria Iron Works, 126, 147
Campbell and Symonds, 39–40
Campbell, John, 22, 36, 38–40, 96
Canned food for the Union Army, 138
Cannonball production, 137
Capitalism, 65
Carbon, and properties of steel, 7
Carey
  Henry, 89
  Henry C., 70–71, 127
  Matthew, 57, 68
Carnegie
  Andrew, vii, 7, 20, 30, 32, 43, 52, 83, 106, 111, 124–125, 142–143, 145, 147, 149–150, 156
  Tom, 143
Carnegie brothers, 144
Carnegie Steel, 68
Carroll, Charles, 16, 113
Cast iron, widely used in colonial times, 6
Cast steel, 135
Catalan furnaces, 11, 100
Celtic iron-making technology, 5
Chalfant, 144
  J. W., 143
Charcoal furnaces, 11, 13, 93, 100, 121
  laborers, 14
  poor construction, 122
Chicago, coal sources, 103

Civil War, 96–97
  demand for iron products, 120
  charcoal furnaces, 5
"Clapboard democracy", 56
"Clapboard junto", 56
Clay Furnace, 107
Clay, Henry, viii, ix, 55–68, 70–71, 78, 80–83, 89–90, 93–94, 98–99, 114–115, 131, 133, 150–152, 157
Cleveland Iron Mining Company, 100
Cleveland, Ohio, 99, 101
  population growth due to pig iron production, 98
Clinton Furnace, 121, 141–142
Coal
  high demand for pig iron production, 106
  importance, 103
  importance to pig iron industry, 103
  necessary for iron and glass production, 103
  necessary for iron production, 106
  replaced wood for home heat and industrial use, 105
  transportation, 106
Coal industry, co-dependency with iron and steel industry, 110
Coal mining
  apprentice system, 105
  child labor, 109
  from craft to cheap labor, 104–105, 109
  early, 103–104
  immigrant labor, 109–111
  wages, 110
Coinage Act of 1873, 82, 84
Coin's Financial School, 84
Coke
  blast furnace development, 121
  coal product, 103
Coke (continued)
  furnace conversion, 120
  needed in puddling furnaces, 106
  properties, 108
  replaced wood in British iron furnaces, 107
Cold Springs Armory, 132
Coleman family, 15
Colliers, 13, 104
Colt, Samuel, 99
"Columbiad" cannon, 134
Conestoga horses, 40
Conestoga wagons, 40
Confederate raids, 128–129
Connecticut, pig iron industry, 132
Connecticut Land Company, 100
Connecticut Valley ironworks, 12
Connellsville, 125
Connellsville coke, 121
Coon, Carleton, 5
Corey families, 104
Cornwall Furnace, 11
Cort, Henry, 49
"Country holes", 105
Cowan, Christopher, 49
Cowley, William, 143
Craig
  Isaac, 36–37, 40–41, 43–46, 74, 79, 105, 159
  Neville B., 79–80
Craig family, 108
Craik, James, 29
Crawford
  Alexander, 93, 158
  Andrew, 160
Cresap, Thomas, 36, 39
Crescent Iron Works, 116
Crimean War, 96, 128
Croghan
  George, 22, 27–30, 39
  Mary, 43
"cupolas", 41

Cyclops, 144
Dahlgren, John, 134
Dalzell, 144
Darby, Abraham, 106–107
Day, William, 160
Democratic Party, 56
Denny
  Ebenezer, 36–37, 46–48, 78
  Elizabeth (O'Hara), 46
  Harmar, 46–47, 79, 81
  Nancy (Wilkins), 46
Denny family, 46–47
Depew, Chauncey, 98
Dickens, Charles, 63
Dinwiddie, Governor, 27–28
"Dumping", 126
  by Britain after the War of 1812, 76–77, 89, 115
  pig iron, 59
  textiles, 59
Duquesne Club, 137, 142–143
Eagle Furnace, 92–93, 141
Eaton, Daniel, 89
Economic depression of 1816-1821, 77
Economy Village, Pennsylvania, 66
Edgar Thomson Works, 145–146
Education, a role of the Presbyterian Church, 32
Edward VII, King of England, 80
Eichbaum
  Peter William, 48–49
  William, 80
Eichbaum family, 81
Eisenhower, Dwight David, 60
Eliza Furnaces, 121, 125, 133, 141
Elizabeth Furnace, 11, 123
Emerson, Ralph, 80
Engels, Friedrich, 94
English furnace workers, 14
Erie Canal, 92
Escanaba Furnace, 146

Evans
  David, 48
  Oliver, 49
Exchange Bank of Pittsburgh, 78
Factory system, 63
Falcon Furnace, 123–124, 141
Farmer, James, 10
Farrell, James, 160
Federalist Party, 73
  split on tariffs and protectionism, 78
Findlay, John, 28
First Presbyterian Church of Pittsburgh, 46, 78, 124
Flinn, William, 150
Floyd, John, 135
Ford, Henry, 20
Fort Hamilton, 136
Fort Pitt, 22, 30, 36, 40, 43, 46, 78
Fort Pitt Foundry, 41, 133–134, 136
Forward integration, 125
Foster
  Stephen, 41, 52
  William, 41
Franklin, Benjamin, 6
Franklin Institute, 64, 107
Fraser, John, 22, 28–29, 39
Free trade, 56, 59–60, 62, 65, 69, 127
Freemasons, Scottish Order, 31
Fremont Furnace, 91
Frick, Henry Clay, 32, 70, 83, 103, 106–107, 110–111, 121, 150, 156, 160
From the Earth to the Moon, 136
Fur trading, 19, 26–27, 39
Furnace workers, nationalities, 14–15
Gage, Thomas, 29
Gallatin, Albert, 74–76
Garfield, James, viii, 152, 156

Index

Gary, Judge, 160
Gates
  Horatio, 29
  John "Bet-a-Million", 160
German ironworkers, 10–11, 14
German settlement in Pennsylvania, 22
Germania ironworks, 10
Germans, supported the Federalist Party, 81
Gibson, John, 36
Gilded Age, 84
Gist, Christopher, 28, 30, 44–45
Glasgow, major trade center, 22
Glass
  Alexander, 137
  Andrew, 137
Glassworks, 43, 75, 105
Gold standard, 82, 84
Goodwin, Lawrence, 155
"Gopher holes", 105
Grace Furnace, 123
Grace I Furnace, 92
Graff, Bennett and Company, 142–143
Grant Hill Iron Works, 50
Grant, Ulysses, 80, 151
Great Britain
  dependent upon Swedish pig iron, 9
  strained relations with Sweden, 10, 12
  tensions with Scots-Irish, 13
Great Exhibition of 1851, 64, 69
Great Lakes iron ore deposits, 4 (see Lake Superior ore)
Greeley, Horace, 80
Greene, Nathanael, 16
"Greenies", 105
Grubbs family, 15
Grubbs, Peter, 11
Gunsmiths, 39
Halket, Sir Peter, 30

Hamilton, Alexander, ix, 19, 31, 55–57, 59, 99, 151
Hanging Rock, 95–97, 128, 137
Hanna
  Marcus Alonzo, 152
  Mark, viii, 83, 85, 155–156, 160
Hanna Furnace, 123
Harding, President, 151
Harpers, 136
Harrison, William Henry, viii, 80, 85, 156
"Harry of the West" Furnace, 90
Harvey, William, 84
Haselton No. 2 Furnace, 141
Hayes, Rutherford, viii, 128, 151, 156
Heaton, James, 90, 160
Hecla Furnace, 96, 128
Heinz, H. J., vii, 70
Hematite ore, 3–4, 122
Henry, Patrick, 19, 61
Hermitage Furnace, 47
Herron, C. B., 143
Highland regiments, 30
High-stack furnaces, 5, 8
Hocking Hills, 87, 95–98, 117, 127, 133–134, 136–137
Holly, Alexander, 99
Hopewell Furnace, 89
Hopkins, John, 43–44
Houston, Sam, 19
Hubbard, Chester, 117
Hughes, Gideon, 90
Huntingdon Turnpike, 79
Hussey, Wells and Company, 135, 138
Industrial Revolution, vii, 7–8, 84, 152
Irish Catholics, suppression by England, 21
Irish labor, 105, 109

Iron
  complex reduction process, 4–5
  difficult to extract, 2
  essential nutrient, 2
  a gift from the gods, 2
  Iliad reference, 2
  important in gaining freedom
    from Britain, viii
  military advantage over copper
    and bronze, 3
  mystical role in many
    civilizations, 2
  Old Testament references, 2–3
  origin and importance, 1
  protection against demons and
    witches, 2
  a symbol of power, vii, viii
Iron Act of 1750, 56, 88
Iron Age, vii, 6
Iron Cross award, mystical roots, 2
"Iron Knight", (see William the
  "Iron Knight" Wilkins)
Iron ores, 3
Iron plantations, 10–12, 15–16, 100
  labor shortages, 10
  in Northern Ireland, 12
Iron production
  in colonial America, 12
  democratizing influence, 5
Iron Prohibition Act of 1750, 9, 16
"Iron Whigs", 82, 87, 90, 98, 127,
  131, 149, 151
Irwin
  Boyce, 48
  James, 16
  John, 36
Isabella Furnace, 143–145
Ishpeming, Michigan, 100
J. B. Corey and Company, 108
Jackson, Andrew, 19, 41, 57, 62–
  63, 65–66, 70, 74, 77–78, 80–82,
  91, 115, 127
Jacksonians, 62–64, 66, 68

James Ward and Company, 94
Jefferson Furnace, 96
Jefferson, Thomas, 20, 47, 50, 55–
  56, 58, 60, 75–76
Jeremiah, Old Testament
  metallurgist, 3
Johnston, John, 38
Johnstown, 126
Jones, 144
  Benjamin Franklin, 123–126,
    143, 150–151, 159–160
  Bill, 146
Jones and Laughlin Steel, 123, 125,
  133, 147
"Jones Mixer", 146
Juniata Rolling Works, 50–52, 114
Kelly
  "Pig Iron", viii, 71, 82–84, 95,
    139, 151–152
  William, 7
Kennedy, Julian, 97, 146
Kidney ore, 4, 89
Kier, Samuel, 123–124, 150
Kirkpatrick, Abraham, 36
Kloman
  Andrew, 144
  Anton (Anthony), 144
Kloman brothers, 146
Kloman, Carnegie and Company,
  144
Kloman-Phipps Iron City Forge,
  144
Knap, Charles, 133
Knox, Philander, 82–83, 160
Krupp Works, 134, 136
La Belle Mill, 116
Lake ore, 93
Lake Superior ore, 92–93, 97, 101,
  115, 117, 121, 123, 137, 141
Lakeville Furnace, 12, 100
Laughlin, 144
  James, 82, 125–126, 151, 159

Lauth
  Bernard, 124–125
  John, 124–125
Lee
  Charles, 29
  Roswell, 63
  Thomas, 28
Leman, Peter, 39
Lewis Dalzell and Company, 143
Lewis, Oliver and Phillips, 136, 142
Lewis, William, 142
Lincoln, 131
  Abraham, viii, 12, 68–71, 78, 80, 98, 125, 127, 129, 137, 142, 150–151
Lodge 45 of Ancient York Masons, 36, 44, 46, 78–79
Loudon, Lord, 30
Lowell
  Francis, 99
  Francis Cabot, 63
Lowell, Massachusetts, textile mills, 69–70
Lower and Upper Union Mills, 144
"Lucifer" twenty-inch cannon, 136
Lucy Furnace, 144–146
Lukens Company, 73
Lyon and Shorb, 137
MacFarlane, James, 31
Machine Age, 99
Machines and mechanics, 64
Mackintosh-Hemphill Company, 41
Madison, James, 19–20, 59, 76
Magee, Christopher, 150
Magnetite, 3
Mahoning Valley, 87, 89, 128, 141
  blast furnace development, 94
  coal mining, 90
  evolution of modern blast furnaces, 93
  first furnace, 90

Mahoning Valley (continued)
  special qualities of its coal, 97
Manufacturers and Mechanics Bank, 115
Marmie, Peter, 16
Martin's Ferry, 117
Marx, Karl, 94
Masonic lodges, 137
Mather, Samuel L., 100–101, 160
Maycock, John, 117–118
McClurg, Joseph, 40–41, 47, 133
McElroy, John, 52
McKinley
  David, 88
  James, 56, 88, 90
  Nancy (Allison), 90
  William, vii, viii, ix, 12, 33, 47, 55–58, 60–61, 63–64, 68–71, 80, 82–85, 87–88, 91, 93–95, 98, 128, 138–139, 151–160
  William Sr., 88, 90
McKinley Tariff of 1890, 84–85, 102, 149, 152–154, 157
Meason family, 15
Meason, Isaac, 36, 44–45, 74, 106, 108, 120–121, 160
Mellon
  Andrew, 160
  Sarah (Negley), 108
  Thomas, 20, 32, 82, 103–104, 106–109, 150
Mellon family, 49
Mendenhall, Cyrus, 117, 121
Menghini, Vincenzo, 2
Merchants and Manufacturers Bank, 52
Miller
  Thomas, 143
  Tom, 144
Mineral Ridge coal deposits, 94
Miners, Welsh and Cornish, 14
Missouri Iron Works, 117
"Moloch" twenty-inch cannon, 136

Monitor, 96, 128
Monongahela House, 80, 127
"Monongahela Rye", 23–24, 30, 74, 76, 142
Monroe, President, 61, 80
Morgan
    Daniel, 29
    George, 36
    J. P., viii, 71, 106, 156
    Jim, 128–129
Morris, Robert, 45
Mount Savage Mill, 67
Mount Vernon Furnace, 45
Nails, 48, 50–52, 106, 116–117, 137
Napoleon, 138
National banking system, 81
National Road, 76–77, 79, 114
National Tariff League of America, 95
Negley
    Alexander, 108
    Sarah, 108
Negley family, 81
Neville
    Amelia, 79
    General, 31
    John, 43
    Presley, 36
Neville family, 108
New England
    manufacturing sector, 58
    textile industry, 64–65, 69
New Harmony, Indiana, 66
New York, pig iron industry, 132
Niles, Hezekiah, 89
Niles, Ohio, 89–91, 102, 123
    birthplace of Wm McKinley, 94
Nimrod Furnace, 123
North Forge, 113
Norton, 137
    Edward M., 116
    George W., 116

Norton, Bailey and Company, 116
O'Hara
    Elizabeth, 46
    James, 22, 36–37, 40–48, 53, 74, 76–78, 81–82, 103, 137, 142, 150, 152, 159
Ohio and Pennsylvania Railroad, 78
Ohio canal system, 93
Ohio Company of Virginia, 28, 39
Ohio's Western Reserve, 99
Oliver, 144
    Henry, 142–143, 151, 160
Oliver Brothers, 143
Onieda, New York, 66
"Original six", 144
Owens, Michael, vii
Paine, Thomas, 20
Painter and Sons, 143
Panic of 1819, 89
Panic of 1837, 93, 115
Panic of 1857, 117
Panic of 1873, 84
Parrott guns, 132
Parrott, Robert, 134
Patron Saints, 2
Penn family, 39, 43
Penn, William, 22, 88
Pennsylvania and Ohio Canal, 92
Pennsylvania Canal, 124
Pennsylvania Railroad, 126, 142
Pennsylvania rifle, 39
Pennsylvania Society for the Encouragement of Manufacture, 70
Pennsylvania Turnpike, 79
Perkins, Jacob, 51
Perry, Commodore, 41
Phipps
    Henry, 143–146, 150–151, 160
    John, 143
Phoenix Company furnace, 67
Phoenix Furnace, 93, 123

Pickands
  Henry, 101
  James, 101
Pickands, Mather and Company, 101–102, 141
Pig iron
  described, 6
  a strategic war material, 16, 41
  Swedish considered the best, 10
  used in Bessemer steel production, 8
Pig Iron Aristocracy
  accomplishments, viii
  decline, ix
  described, viii
  dominated by Scots-Irish, 19
  lacked a political party until the 1850s, 79
  marriage partnerships, 37
  military members, 36
  nationalities, 19
  Ohio, 88
  paved the way for the steel barons, ix
  philanthropy, 150
  political influence, viii
  Republican Party relationship, 37
  responsible for electing Republican leadership, 98
  slowly converted to steel masters, 147
  symbiotic alliance with railroads, 126
  U.S. Army connections, 36
"Pig-Metal Aristocracy", 98
"Pigs", 6, 14
Pilgrims, 99
  built ironworks, 9
Pitcairn, Robert, 142–143
Pittsburgh
  coal sources, 103
  political importance, 80
Pittsburgh Coal Seam, 41, 104
Pittsburgh Foundry, 41, 47
Pittsburgh Gazette newspaper, 79–80
Pittsburgh Library Company, 37
Pittsburgh Wheeling Steel, 117
Political bosses, 150–151
Politics, 46
  nationalism versus regional, 59
Pontiac, 29
Powderly, Terence, 156
Presbyterian Church, 31–32, 37–38, 137
Principio Company, 113, 116–117
Principio Creek iron furnace, 10–11
Protectionism, 55–57, 62, 65, 68–70, 75, 77, 79–80, 83, 126, 129, 131, 133, 137, 149–151
Puddler furnace cinder, 146
Puddling, 49–50, 105
  first use by Isaac Meason, 45
Pullman, George, 70
"Pumpkin rollers", 105
Puritans, 99–100
  built ironworks, 9
Putt family, 15
Quakers, 22, 151
Quay, Matthew, viii
Railroad rails, manufacture, 67, 120, 147
Rebecca Furnace, 90
Report on Manufactures, 57
Republic Steel, 123
"Republican" Party, 56
Republican Party, 37, 64, 75, 78, 81, 83, 126, 133, 137–138, 142, 149–151
  formation, viii
  (Jeffersonian Democrats or Democratic-Republicans), 73, 75–77, 81
Revere, Paul, 16, 100

Robinson, William, 78
Rockefeller, John D., 152
Rodman, Thomas Jefferson, 133–135
Rogers, Mahlon, 49
Rolling mills, 11
Room Number 6, 142–143
Roosevelt, Theodore, 80, 83, 160
Ross
   George, 11, 16
   James, viii, 77–78
Russell, Thomas, 91, 113
Saint Clair
   Arthur, 16, 36, 42, 44, 47, 74, 78
   John, 29
Saint Clement (Pope Clement I), general patron of blacksmiths, 2
Saint Dunstan, patron of farriers, 2
Saint Eloi, 2
Saint Leonard, patron saint of prisoners, 2
Saint Quentin, 2
Salt production, 106
"Satan" twenty-inch cannon, 136
Saugus, American's first large industrial complex, 9
Scaife family, 49
Scaife, Jeffery, 49
Schenley, Captain, 43
Schenley Park (Pittsburgh), 43
Schwab, Charles, 146
Scientific American, 136–137
Scots-Irish
   charcoal furnaces, 23
   cooperation with frontier Indians, 26
   domination of Pig Iron Aristocracy, 19–21
   founding fathers of Pittsburgh, 36
   hatred of British, 19
   infuriated by iron and whiskey tariffs, 56

Scots-Irish (continued)
   linguistic legacy in western Pennsylvania, 24
   opened the first colonial forges and furnaces, viii
   political power, 27
   recruited by William Penn, 22
   regiments, 29
   in the Revolutionary War, 31
   symbiotic relationship with German settlers, 26, 38
   whiskey production, 23
Scots-Irish furnace workers, 14
Scott, Thomas, 126
Shaker utopian manufacturing communities, 66
Sheet mill, 49
Sheridan, Philip, 80
Sherman
   Senator, 152
   William, 80
Sherman Silver Purchase Act, 85
Shipbuilding, in Pittsburgh, 42
Shirley, William, 30
Shoenberger, 137
   George, 114
   John, 52–53, 115, 151
   Peter, 51–52, 101, 114–116, 159
Shoenberger and Company, 52
Shoenberger, Blair and Company, 133
Shoenberger family, 15
Singer, Nimick and Company, 135
Slavery, on northern iron plantations, 12, 14
Sligo Rolling Mill, 50
Smith
   Adam, 20, 65, 69, 127
   James, 143
Smith Foundry, 143
Spang, 144
   C. H., 143
Spang, Chalfant and Company, 143

"Splint" coal, 97, 107–108
Spotswood, Alexander, 10
Spotsylvania ironworks, 10
Springfield Armory, 63
Stackhouse, Mark, 49
Stackpole and Whiting nail mill, 51
Stackpole, William, 51
"Steam coal", 114
Steam engines, replaced water-driven bellows, 122
Steel, 6–7
   preceded by pig iron, viii
Steel Age, 156
Steel industry, blossomed due to tariffs, 68
Stephen, Adam, 29
Stephens, Edward W., 80, 115–116
Stephens family, 81
Stevens, 160
   E., 151
   Thaddeus, 71, 81
Stewart, Robert, 50
Struthers Furnace, 144
Struthers, Robert, 89, 159
Superior Furnaces, 133
"Swamp Angel" cannon, 96, 128
Symington, John, 135
Symonds, John, 39
Taft, William Howard, 91, 139, 151, 158, 160
"Tariff of Abominations", 62, 66
Tariff of 1816, 59–60
Tariff of 1824, 61–62, 64, 89
Tariff of 1832, 67, 91
Tariff of 1842, 67–68, 95, 115
Tariff of 1862, 131–132
Tariff of 1890, viii (*see* McKinley Tariff of 1890)
Tariffs, 56, 59, 64, 70, 78
"Task system", 14
Tatara, Japanese pig iron reduction method, 5
Taylor, Zachary, 80

Tecumseh, 29
Temperance Furnace, 90
Thaw
   Harry, 151
   John, 74, 138
   William, 124, 138, 150–151, 160
The Harmony of Interests, 70, 127
The Wealth of Nations, 20, 65
Thomson, Edgar, 126
Tin plate mill, 49
Tinplate, 11, 102, 138, 152–155, 158
Tod, David, 90–93, 160
Top Mill, 114–115
Tubalcain, Old Testament blacksmith, 3
Turnbull
   John, 45
   William, 16, 44
Turnbull and Marmie Company, 16, 45
Tuyeres, 121
Twain, Mark, 80
Tyler, John, 81, 94
Union Furnace, 45, 96
Union Iron Mills, 143–144, 146
Union Rolling Mill, 49–50
United States Bank, 74
United States Steel, 146, 156, 158
Utopian manufacturing communities, 65–66
Verne, Jules, 136
Vertical integration, 52, 125
Vesta Mines, 125
Victoria, Queen of England, 69
Virginia Mill, 116
Wako, steel used in Samurai swords, 5
Wallace, George, 48
War of 1812, 76, 89
   economic consequences, 58–59

Ward
　James, 94, 123, 160
　James Sr., 91
　William, 91
Warner, Jonathan, 92, 158, 160
"Warner's Scotch Pig", high quality pig iron, 92
Washington
　Augustine, 11
　George, 11–12, 16–17, 22–23, 28–31, 35, 42, 74, 80, 88, 99
　Lawrence, 28, 88
Watertown Arsenal, 133
Wayne, Anthony, 41–42, 45
Ways and Means Committee, 58, 83, 95, 138, 151–152
Webster, Daniel, 60–61, 70
Welsh furnace workers, 14
West Point Foundry, 133–134
Western Reserve, 101, 151
Westinghouse, George, vii, 32, 70, 99, 135, 160
"Wheats", 105
Wheeling Iron Works, 114
Wheeling Steel Company, 117
Wheeling, West Virginia, 137
　coal sources, 103
　expansion of pig iron production, 113
Whig Party, 60, 64, 68, 75, 78–79, 81–82, 92, 94, 115, 133
　formation, viii

Whiskey production, 15, 23
Whiskey Rebellion, 25, 31, 42, 56, 75–78, 89, 131, 142
Whitaker
　George, 113, 116–117, 159
　Joseph, 116
　Nelson, 117
Whitaker family, 113, 117
Whiting, Ruggles, 51
Whitney, Eli, 63, 99
Whitwell, Mr., 145–146
Wilkins
　John, 36–38, 46, 74
　John Jr., 46
　Judge, 57
　Nancy, 46
　William the "Iron Knight", 46, 57, 61, 77–78, 80–81
Wilkins family, 46–47
Wilson, James, 143
"Winter diggers", 105
Winthrop, John Jr., 9
Wire mill, 49
Wolfe, General, 30
Wrought iron, 6, 105
Wrought iron mills, 50
Yellow Creek, Ohio, 89
　first furnace in Mahoning Valley, 87
Youngstown, Ohio, 89, 123, 141
　coal sources, 103
Zoar, Ohio, 66

www.ingramcontent.com/pod-product-compliance
Lightning Source LLC
Chambersburg PA
CBHW071425160426
43195CB00013B/1810